Livestock and Livelihoods
The Indian Context

Environment & Development Series

Sustainable Development in Fragile Environments, *N.S. Jodha*

Sustainable Technology: Making the Sardar Sarovar Project Viable, *Suhas Paranjape* and *K.J. Joy*

Waste Minimization: A Practical Guide to a Cleaner Production and Enhanced Profitability, *Prasad Modak*

Joint Forest Management: The Haryana Experience, *Madhu Sarin*

Banking on Biomass: A New Strategy for Sustainable Prosperity Based on Renewable Energy and Dispersed Industrialization, *K.R. Datye*

Minding our Marine Wealth: An Appraisal of Gujarat's Coastal Resources, *N.D. Chhaya*

Nurturing Biodiversity: An Indian Agenda, *Madhav Gadgil* and *P.R. Seshagiri Rao*

Renewable Energy and Environment: A Policy Analysis for India, *N.H. Ravindranath, K. Usha Rao, B. Natarajan and P. Monga*

Towards an Agroecosystem Policy for India: Lessons from Two Case studies, *A. Damodaran*

Water Harvesting & Sustainable supply in India, *R. N. Athavale*

Livestock and Livelihoods
The Indian Context

Dr Nitya Sambamurthi Ghotge

Centre for Environment Education
Ahmedabad

FOUNDATION
BOOKS

Delhi • Bangalore • Mumbai • Kolkata • Chennai • Hyderabad

Published by
Foundation Books
4764/2A, 23 Ansari Road, Daryaganj, **New Delhi** - 110 002.

FOUNDATION BOOKS
C-22, C-Block, Brigade M.M., K. R. Road, Jayanagar,
Bangalore - 560 082.
Plot No. 80 Service Industries, Shirvane, Sector-1, Nerul,
Navi Mumbai - 400 706
60, Dr. Sundari Mohan Avenue, 1st Floor, **Kolkata - 700 014**
21/1 (New No. 49), 1st Floor, Model School Road, Thousand Lights,
Chennai - 600 006
House No. 3-5-874/6,4, (Near Apollo Hospital), Hyderguda,
Hyderabad - 500 029.

The **Environment and Development Book Series Project** is funded by the Swiss Agency for Development and Cooperation and managed by the Centre for Environment Education (CEE). CEE is a national institute, established in 1984, and supported by the Government of India, Ministry of Environment and Forests.

Project Committee	:	Kartikeya V. Sarabhai, Seema Bhatt, Kiran Desai, Prithi Nambiar, Bipin Shah
Series Editor	:	Prithi Nambiar
Production Assistance	:	Maitri Baruah, Snehal Bhatt, Snehalatha S.V.
Editorial Assistance	:	Saira Lakshminarayanan
Layout	:	Balamani R. Menon
Photographs	:	Shri Dilip Sinhji, Shri Chanakya Sinhji, Dr Nitya Ghotge, Shri Vivek Broome and ANTHRA photo files
Cover Design	:	S. Rajkumar

Jointly published by Foundation Books and the Centre for Environment Education

ISBN 81-7596-183-X
Published by Manas Saikia for Foundation Books and Printed at Print Perfect, Mayapuri, New Delhi.

Foreword

The Centre for Environment Education is concerned with bringing more innovative strategy to the notice of policy-makers and the general public in order to sensitize them to the need for environmentally sound decisions. The current book series has been planned as a means of providing access to alternative viewpoints, backed by analysis and information, on various policy aspects presented by development theorists and practioners from diverse fields.

The relationship between humans and other species valuable for various purposes extends beyond pure economics into the realm of compassion and cooperation, as we are here in the role of chief caretakers on the planet. Human habitation demands the use of draught animals in farming, dairy operations, milking, transport and others. The agrarian economy demands the optimum utilization of species such as the camel, goat, mithun, yak, cow, buffalo and horse to take care of and assist in many daily activities around which our well being is centered.

Caring for these animals goes beyond providing veterinary services or enlisting them in shelters when they are no longer useful to us. Even royalty in the past has made use of horses, elephants and other animals as a mark of regal splendour and the Indian tradition of cattle fairs and shadies is replete with instances of mutual benefit-sharing between our own species and these others. Buddhism and Jainism too have contributed to this tradition of kindness and compassion to animals. It is essential therefore that a book, which explores the many facets of interrelationship between humans and other animals, be brought out in order that the existing resources of biodiversity and livelihoods do not dwindle out of neglect. To do this, it is essential to consider aspects of animal history such as terrain diet, herding practices, medicine and others that ensure their health and continued utility so that a holistic perspective of the transition from a pastoral economy to agrarian and industrial-based values can be made.

The ideas presented in this book are based on practical fieldwork carried out by a number of NGOs in veterinarian care and animal health and nutrition. Therefore, along with the work done by grassroots organizations, many of which, work for the rural people using their indigenous systems of knowledge, in particular, the rich heritage of Ayurveda and Unani texts, it is hoped that the book will pave the way to future subjects of research that will generate an ethno-veterinary, a pro-ecological niche for sciences declining with problems of transition. The book also studies such related activities as milking, calving, feeding, herding and rearing cattle, in order to examine possibilities of their interaction both domestically and as contributors to the national economy.

Since clinical veterinary medicine and surgery are areas of research, which deal with livestock development and indigenous knowledge systems, there is a perceived need to plan for gender-sensitive and sustainable alternatives while training para-veterinarians, designing educational material and developing areas of interest relevant to our villages that thrive on an agrarian economy. This book, therefore, deals with these issues in addition to promoting methods of sustaining biodiversity, producing alternative systems of veterinary care and other development related areas such as women's role in this economy. The framework for analysis presented in this book should also help policy-makers and NGOs working in related fields of biomedicine choose viable and sustainable alternatives to the stresses imposed by modern living.

An overall picture would necessarily involve the educational infrastructure needed, advocacy on key policy issues and a grassroots training methodology to assess the crossover from rural based to machine based economies. It is hoped that this book will provide the necessary background for further studies in this new and exciting cross-frontier discipline.

Kartikeya V. Sarabhai
Director
Centre for Environment Education

Acknowledgements

There are many who have contributed to the writing of this book.

I am especially grateful to:

Shri K.V. Sarabhai, Prithi Nambiar and the CEE team for inviting me to write the book.

My farmer and livestock rearing friends for whom this book is written. Friends from Tarwal, Andgaon, Janapur, Harpud Ginabahar, Dediapada, Bhimavaram, V. Kota, Havanur and many other villages across India .

Mr G. Madhusadan Pillai, Former Animal Husbandry Commissioner, Maharashtra State and presently Director General Maharashtra Energy Development Agency, for encouraging me to put my ideas and thoughts on paper as well as supplying me with books, reference material, papers and documents.

Dr Niranjan Mehta for showing me that present systems of livestock rearing are "Far from Paradise" and Robert Paget for showing me a little glimpse of "paradise" as well as organic /biological farming systems for animals.

My long-suffering and extremely tolerant family, my parents, Uma and Sam, my husband Sanjeev and my children Nayantara and Rishabh for putting up with a semi-migratory, nomadic family member.

Dr Jayvir Anjaria, Dr Usha Mandokhot and Dr Vallabh Mandokoth for their support.

Sagari Ramdas, friend, partner and fellow traveler: Much of this book is drawn from experiences we have shared together.

My colleagues at ANTHRA especially Ms Asha Ogale for help with the index and Anito Antoni for help with that machine which has come to rule our lives, "the computer".

Ilse Kohler Rollefson for information especially on Rajasthan and camels.

Shri Dileep Sinhji and Shri Chanakyaji of Ahmedabad for excellent photographs of the Kankrej and the Banne Buffalo.

Sanjeev Ghotge, besides being an extremely tolerant and supportive member of the family, debated and discussed issues, provided essential "fodder for thought", references where I thought there were none as well as read and edited the manuscript.

To all ....my sincere thanks

Dr Nitya Sambamurti Ghotge

Contents

Preface

They came to the Indian subcontinent in search of pastures for their animals. Across the cold and rugged mountain ranges, they herded their cattle and their horses and started a civilization which was to have an everlasting impact on the Indian sub continent. The Aryans, who were primarily cattle rearers, found pasture and settlements on the banks of the Himalayan rivers. The favourite food of the Aryans was *panch amruth* or five nectars based on products derived from milk: ghee, yoghurt and milk itself flavored with honey. Abundance was symbolized by the divine cows...Kamadhenu, Nandini and Surabhi, gifted by the Gods to the sages. The favourite god was a cowherd or Gopal whose descendants the Yadavs still practise a profession started by their ancestors thousands of years ago.

The importance of livestock in the story of human civilization is significant especially with reference to the Indian context. Animals continue to cater to the ever-increasing needs of a single species, the *Homo sapien*. Livestock and livestock products contribute in some way to every household, the contribution often increasing as one goes down the economic ladder. Livestock rearing forms an important livelihood activity for most Indian farmers supplementing incomes, offering employment opportunities, supporting agriculture as well as contributing to the health and nutrition of millions of households.

The livestock sector accounts for 9 % of the GDP next only to agriculture in the rural sector. However, this sector seldom performs at optimal levels and is riddled with inconsistencies and contradictions. Although India today has the distinction of having attained the status of being the largest producer of milk in the world, the average rural household in most of the country still cannot afford to consume milk. We are the original home of some of the finest breeds of cattle but these breeds are fast disappearing in the country and the best germplasm of these breeds are found in countries as distant as Australia and Brazil. We have ancient texts eliciting in detail the health and care of cattle, elephants and horses but animal management and health remain major issues in the rural areas.

History is said to be a dialogue of the present with the past. It becomes even more important to look into the history of livestock in our country as we step into a new millennium and peer into an uncertain future. Perhaps the past will help us understand our inconsistencies and contradictions better and thus pave the way for better programmes and policies which do not constantly defeat each other.

Thematically, this book is divided into three parts. The first part deals with the history of domestication and traditional systems of animal rearing. The second part deals with emerging systems and the contradictions within as well as future global trends. The third discusses possible alternatives. Each of the themes itself is so large that it has not been possible to cover the entire range of subjects within a topic. However, an attempt has been made to link different issues together and present the mosaic perhaps a little more clearly.

One has tried to be dispassionate while presenting the facts but sometimes passion forces us to speak a little louder and a little more harshly, especially when policies and programmes are designed without a care or concern except the economic one which then forces animals to be used and exploited for human needs and then abandoned when they are no more useful. Let us not forget that the care of another species requires compassion and the relationship between domestic animals and their human caretakers is a very old and compassionate one.

1

In the Beginning...

The Domestication of Animals

The story of the domestication of animals is over 14 thousand years old. It is believed that the first animals to be domesticated were dogs, presumably as companions to the human race during hunting. Through the years, different species have been domesticated in different parts of the world from their wild progenitors to cater to different human needs of food, fibre, fuel, medicine and transport. In some cases these animals have spread throughout the world. For example, cattle, goats and sheep. In others, they have remained localized in their region of origin like the alpaca, llama, and yaks. Yet others are in a state of semi-domestication like the elephant, *mithun*, mink and silver fox.

Table 1 The Domestication of Important Livestock Species

	Species	Latin name	Wild pro-genitor	Latin name	First known domestication progenitor		Distribution of wild
					Date	Place	
1.	Water buffalo	*Bubalus bubalis*	Water Buffalo	*Bubalus arnae*		China and Indo-China	India, S. Asia, possibly West Asia
2.	Cattle (*taurine*)	*Bos taurus*	Aurochs	*Bos primi genus*	6200 BC	Turkey	Europe, Asia, N. Africa
3.	Cattle (*zebu*)	*Bos indicus*	Derived from *Bos taurus*			Balu-chistan, Central Asia	Asia, N. Africa

(contd.)

(Table 1 contd.)

	Species	Latin name	Wild pro-genitor	Latin name	First known domestication		Distribution of wild progenitor
					Date	Place	
4.	Yak	*Bos grunniens*	Yak *mutus*	*Bos* known	Not known	Not Himalayas	Tibet
5.	*Mithun*	*Bos frontalis*	Gaur	*Bos gaurus*	2500 BC		South and SE Asia
6.	Goat	*Capra hircus*	Wild goat	*Capra aegragrus*	7000 - 8000 BC	West Asia	West Asia
7.	Sheep	*Ovis aries*	Mouflon	*Ovis orientalis*	7000 - 8000 BC	West Asia	West Asia
8.	Asian elephant	*Elephas maximus*					S Asia
9.	Drome-dary *darius*	*Camelus drome-*	Drome-dary	*Camelus sp*	3000 BC	W Asia N Africa	Asia, probably
10.	Horse	*Equus caballus*	Wild Horse	*Equus ferus*	3500 BC	S Ukraine	Russia, Central Asia
11.	Donkey	*Equus Asinus*	African ass	*Equus africanus*	4000 BC	Egypt	N Africa, possible W Asia
12.	Pig	*Sus domes-ticus*	Wild boar	*Sus scrofa*	7000 BC	Asia	Europe, Asia and Africa

(contd.)

(Table 1 contd.)

	Species	Latin name	Wild pro-genitor	Latin name	First known domestication		Distribution of wild progenitor
					Date	Place	
13.	Chicken	*Gallus gallus*	Red jungle fowl Ceylon jungle fowl Grey jungle fowl Green jungle fowl	*Gallus gallus* *Gallus lafayetti* *Gallus sonner-atii* *Gallus varius*	4000 BC	South and SE Asia	South and SE Asia
14.	Pigeon	*Columba livia*			3000 BC		Europe, Africa, Asia from India to Japan

Source: Global Biodiversity: Status of the Earth's living resources; Groombridge Brian: Editor, (1992). Compiled by World Conservation Monitoring Centre. London, Chapman and Hall.

The region of domestication and the need for domestication have given rise to certain distinct patterns of livestock farming. A large number of animals were domesticated around the grasslands of West and Central Asia. These were mainly from the herbivorous species that ate grass: sheep, goats, cattle, horses and camels. Early cattle, sheep and goat rearers were migratory and herded their animals in search of grasslands. When the pressure on grasslands became too much, they migrated in search of fresh pastures or invaded fresh territory. Many major invasions including those of the Mongols are attributed to times when the pressure on grasslands became excessive.

In contrast, species of the forest like the elephant, the water buffalo, pigs and chicken were first domesticated in the region of South Asia and Indo-China in areas which were heavily forested and where plenty of water was available. They were domesticated by different forest-dwelling communities for food and draught purposes. Interestingly, these patterns of livestock rearing are still seen today. In the drier parts of the world herders still herd their cattle, sheep and goat in search of pasture while in the forests of the world, different tribes and groups maintain buffalo, pigs and chicken much as their ancestors did thousands of years ago.

The major revolution in livestock farming happened when herding was successfully incorporated with crop farming wherein the wastes from agriculture crop residue and straw could be used to feed animals. In exchange, the energy of animals could be used to work the land and the waste of animals, dung, could be used to fertilise the fields. The other products: grain, seeds, milk, meat and wool, were used for human consumption. It was this great revolution that led to food surpluses and helped societies go beyond the level of mere subsistence.

Animals also came to represent wealth, power and domination. Elephants, horses and camels were used to symbolize the might of the ruler as well as assist preparations in war.

India today represents the entire diversity from tribal communities who are largely dependent on the forests for their survival, and rear chicken mainly for cock fights to a highly organized and fast growing commercial poultry industry, from semi-nomadic pastoralists who annually migrate in search of fodder, to settled dairy farmers who use the latest technology in their units.

But before we discuss this in detail, let us return to the history of domestication and breeds found in India.

2

Patterns of Livestock Rearing

India has the distinction not merely of having a rich and diverse number of animal species, but we also have the distinction of having some of the widest range of breeds within each species, thereby representing a significant percentage of the world's domesticated livestock diversity. Of the breeds found in the world, those found in India are represented below in Table 2.

Table 2 Livestock Breeds Found in India

Species	World	India (FAO)	India (ICAR)
Cattle	787	30	30
Buffalo	72	19	15
Sheep	920	59	42
Goat	351	29	20
Pig	353	3	3
Ass	77	3	3
Horse	384	9	6
Camel	56		8
Chicken	606		18

Source: Sahai, R. and Vijh, R.K. (Eds.). 2000. *Domestic Animal Diversity Conservation and Sustainable Development*. Karnal: SI Publications

This rich diversity is directly accounted for by the fact that the subcontinent itself is diverse in culture, in its agro-ecological zones, in social and economic practices. These breeds have developed and evolved over time to suit the different production goals and requirements of these different farmers.

Traditionally, livestock rearing was also distinctly related to castes and communities with specialized groups rearing distinct animal types. Even today these communities continue to rear these animals and therefore any development programme has to take into account these regional differences, preferences and choices.

Fig 2.1 Major Agro-ecological Zones in India

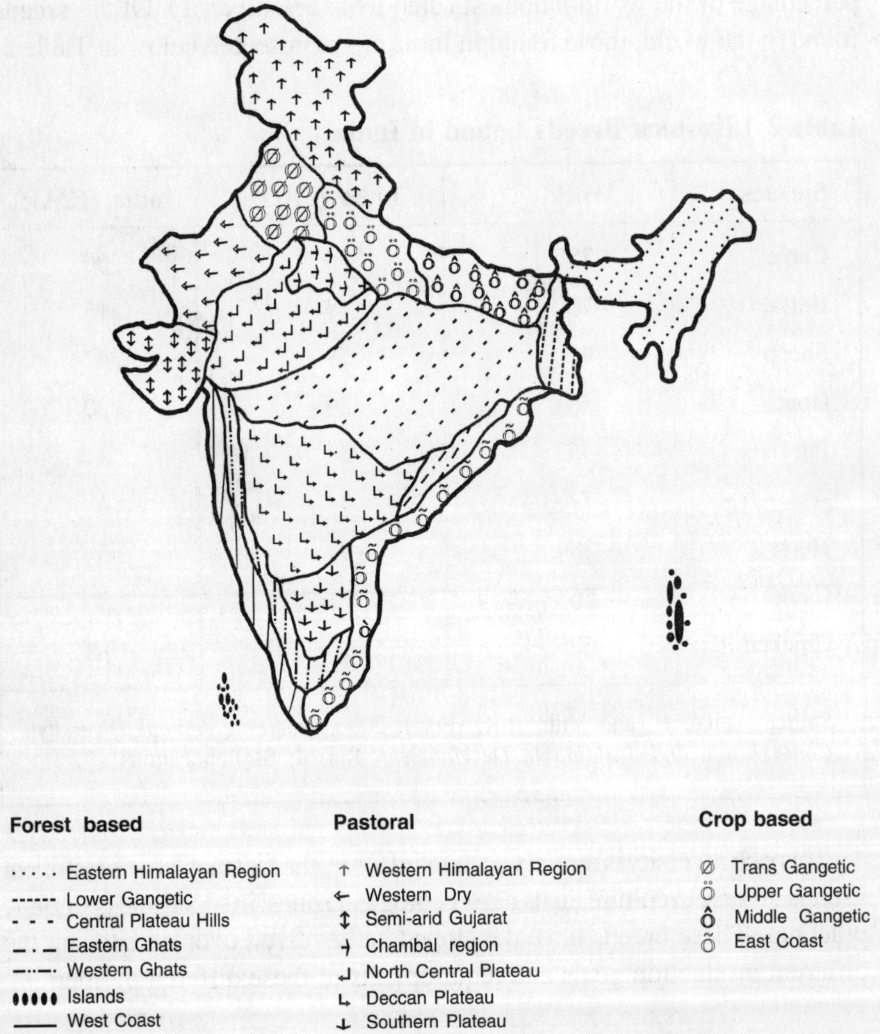

Major Agro-ecological Zones in the Country and Livestock Practices within Pastoral Systems

Large tracts of the country are semi-arid or arid and frequently threatened by droughts. Sustained agriculture is impossible or extremely difficult. However, these regions border grasslands that are covered by pasture over parts of the year. These are the regions where even today, migratory or semi-migratory systems of livestock rearing continue to be practiced. The species most suitable for migrating is the sheep. Sheep live in groups and herds and are primarily grazers as far as their eating habits. Their major food is grass. Often cattle, camel, goat, yak and even buffalo migrate for food.

There are many reasons why pastoralism still exists. Semi-arid regions, grasslands and the arid regions where these systems are practised are ecologically fragile zones that cannot be exploited or abused indiscriminately. Contrary to popular perception, livestock contributes to the wealth of these grasslands by dispersing valuable grass seeds, by keeping unnecessary weeds in check and by fertilizing these grasslands with their dung. In return they consume grass, which cannot otherwise be consumed by humans and convert them to a range of valuable animal products: milk, meat, wool, manure. People living in these areas are almost completely dependent on their livestock resources for their survival because any other form of farming is very difficult.

Pastoral systems are some of the most flexible systems. Livestock numbers increase in response to favourable conditions and shrink when conditions are not favourable. During drought years when grazing lands get eroded or simply when the pressures in the area become too high, pastoral communities have the option of either selling their stock, changing to another less demanding species or migrating to another area where fodder is available in greater quantity. Over the years, pastoral systems have grown, evolved, and adapted to changes in their environment and today many interesting and complex systems are found dotted over the subcontinent.

Some have adopted agriculture and practise crop farming for a few months of the year during the monsoon. The communities who practice pastoralism have developed breeds to suit the conditions and requirements of their region. For example, among sheep, breeds from the Himalayas are wool-yielding breeds, while the breeds from the hot desert are dairy breeds that also yield coarse carpet wool and those in the south are primarily meat breeds and do not yield wool at all. Pastoral systems become non-viable when so called 'development' takes over and the conflicting interests of settled farmers and urban pressures come in the way of migrations.

Major Pastoral Systems in India

The western Himalayan zone

These mountainous regions that form the northern border of the country experience temperate climates. The areas of Kashmir, Ladakh, Himachal and Uttaranchal fall in this region. Land holdings here are small and farming at best difficult because of the extremely harsh conditions in winter. However, in the summer these hill slopes are covered with excellent pasture and are home to a number of pastoral sheep and goat rearing communities the Bhakarwals, Gaddi and Gujjars who migrate into these pastures in the summer and come down into the valley in the winter. Some of the finest wool breeds of sheep are found in this region, such as *Changthang*, *Gurez*, *Karnal*, *Bhakarwal*, *Biangir Bhadarwal* and *Gaddi*. This is also the region of the *Pashmina* goat. Other important breeds of goat are the *Chegu*, *Kashmiri*, *Gaddi*, *Changthangi* and the *Bakharwal*. Yaks are reared in Ladakh and the two-humped Bactrian camel maybe found to a very limited extent in the area bordering Tibet. Slightly south, in the hills of Uttaranchal the Van Gujjars herd buffalo under migratory systems.

Transgangetic plain

Lying in the shadow of the Himalayas, this area covers the states of Punjab, Haryana and Western Uttar Pradesh. While parts are transected by Himalayan rivers, other parts remain semi-arid because rainfall fluctuates greatly between the plains and the sub-montane areas as indicated by the *Report of the Irrigation Commission* (Vol. 2) released by the Government of India in 1972.

In the semi-arid region, sheep, the *Beetal* breeds of camel and cattle are reared under semi-migratory systems, while the area which is irrigated and under cropland is home to some of the finest dairy buffalo breeds of the country like the *Murrah* and the *Nili-Ravi* The predominant crop in this region is wheat. Today these states are heavily irrigated and constitute the epicenter of the country's 'green revolution'. Herders migrate seasonally from the semi-arid areas to the irrigated areas in search of crop residue, which is then exchanged for dung.

The western dry desert

South of the Indus valley and to the west of the Gangetic plains lies the arid region of the country. This region is characterized by extremely hot desert conditions. Agriculture is limited to a few millets and pulses and it is not possible for families to sustain themselves on agriculture alone. Land -holdings are immense but largely unproductive. The main occupation is livestock rearing. Although apparently barren, various fodder grasses and trees species nourish and sustain animals. Two particularly remarkable species are the Sevan grass (*Lasirus indicus)* and the Khejri tree (*Prosopis cineraria)* which support animals through long periods of drought and water scarcity. It is quite amazing that some of the best dairy breeds of cattle and goat originate from this region. The famous cattle breeds of this region are the *Rathi* and *Tharparkar*. A well-known goat breed is the *Marwari* goat. A large number of sheep breeds are found in the region: the *Marwari*, *Jaisalmeri*, *Pugal*, and *Nali*. These breeds produce a coarse wool too rough for garments but suitable for carpets. Livestock herders manage their herds under extensive range systems and migrate during the winter

Box - 1

In Search of the *Rathi*

As the sun descended into the clear desert horizon, our jeeps sped silently along a superb road on the left bank of the Ganga Canal, north of Bikaner city in Rajasthan towards the Pakistan border. The landscape was stark with only a few trees, some camels and almost no human inhabitation.

(contd.)

(Box 1 contd.)

A cool winter wind made the drive extremely pleasant as we made our way to a tiny settlement about 40 Kms from the border. We were in search of the Rathi, a breed of cow which is supposedly a mix of the *Sahiwal,* the *Red, Sindhi,* the *Tharparkar* and the *Dhanni* breeds and their keepers, the Raths, a tribe of Muslim cattle herders.

On reaching our destination we found a small settlement, which consisted of a house and a huge cattle shed. It was empty save for the women of the household. The women informed us that the cows were still away grazing, but were expected to be back soon. They asked us to wait and quietly disappeared. They were not comfortable talking to a group of urban strangers.

Shortly afterwards in the twilight we heard the low tinkling of cow bells and even as we turned we saw the herd returning walking with a dignity that any model would find it hard to match. This gorgeous herd of 40 or more cattle, more disciplined than school children happily wended their way to the shed. One of the women came out and called out softly. A cow came up to her and she placed a vessel beneath it. Very soon we heard the sound of milk being drawn into a vessel for our tea. There was no tethering of the animals, no coaxing and cajoling was required. After being milked, the cow quietly joined the rest of the herd in the shed. Soon we could hear the cheerful sounds of animals consuming their daily ration of special green fodder and concentrate.

There was no doubt that these people and their herds were some of the best managed anywhere in the country. We turned our attention to the Rath who owned the herd. He knew about the best fodder, the best methods of rearing, the best medicines and the kind of methods to be followed when the animals became sick. There was a discernable deep bond strengthened over centuries between this tribe and their herd, which as we realized was not just about adequate management skills, but is also indicative of the ideal kind of relationship humans and animals ought to have.

and summer and normally return to their home villages in the monsoon. Annual migrations usually take them to the agricultural regions surrounding them in Punjab and Haryana. In bad years when there is a severe drought these herders may migrate as far as the states of Madhya Pradesh and Maharashtra in the south. The desert of course is also home to the camel.

The desert is fringed by grasslands like the Banni of the Kutch. Some of the famous grasses of these grasslands are *Sporobolus pallidus, Sporobolus marginatus, Sporobolus helvolus Dicanthium annulatum (Cincotta* and *Pangare*). The *Kutchi* breeds of sheep and goat are from this region. The Banni grasslands have been traditionally used to fatten animals. Farmers from surrounding districts would send their animals to be raised on the special grasses here. Traditionally, Maldhari herders would raise *Kankrej* bullocks in the Banni to sell in irrigated areas of South Gujarat.

Gujarat hill and plain and semi-arid Rajasthan

This region borders the desert and is characterized by low hills and dry thorny scrub forests. In the absence of irrigation, agriculture is limited and livestock is a necessary means of livelihood. This region is home to nomadic communities like the Maldharis, Bhadawars and the Rebaris. Camel, cattle, buffalo, sheep and goat are predominantly reared in this region. Maldhari herders from the Gir forest farm the internationally famous *Gir* breed of cattle and the *Jaffarbadi* breed of buffalo. Breeds of goat in this area are the *Sirohi*, the *Jahkrana* which is now supposed to be threatened (*Jhingua*), the *Mehsana*, the *Zalawadi*, *Gohilwadi* and the sheep breeds are the *Sonadi*, *Malpura* and *Patanwadi*.

The north central hill

This is the region lying south and southeast of the Transgangetic plain. Highly degraded, this area is dissected by landforms known as ravines, an effect of water erosion. The area is drained by the Chambal River and is also known as the Chambal Valley. The vegetation consists of thorny scrub and agriculture is difficult. The species, which can best survive this kind of degraded environment, is the goat, as it consumes almost anything and can also negotiate the ravines easily. The world famous *Jamnapari* and *Barbari* goats are from this region.

The Deccan plateau

Further south is the Deccan plateau, which covers the semi-arid parts of the states of Maharashtra, Karnataka and Andhra Pradesh. Agriculture here is characterized by sorghum, pearl millet, some wheat, cotton, oil seeds, groundnuts and different pulses. The shortage of water allows only one monsoon harvest and agriculture needs to be supplemented with live stock rearing. Shepherding communities in this region such as the Dhangars of Maharashtra, the Kurumas of Andhra and the Kurubas of Karnataka practise semi-nomadic systems of animal rearing. They migrate to the areas with higher rainfall in search of fodder in the winter and summer and return to their homes only in the monsoons where they practise crop farming for three months. These migrations benefit crop farmers in the high rainfall areas, as they invite these shepherds and their animals to camp in their fields for a few days every year. In return for the crop stubble, grain and some times money, the sheep fertilizes the fields of crop farmers with sheep dung, which is highly valued organic manure. The breed of sheep favoured by these communities is the Deccani. Most shepherds also like to rear a few goats and the *Osmanabadi* is the preferred breed. The wool of the Deccani sheep is woven into coarse blankets and is also very popular in the region.

Southern plateau

The semi-arid plateau extends further south through Karnataka and into the state of Tamil Nadu. Here too shepherds practise semi-migratory systems of animal rearing. Sheep breeds like the *Bellary*, *Hassan*, *Mandya* and *Bannur* are popular in Karnataka, while in Tamil Nadu, the *Mecheri*, *Kilakarsal*, *Vembur*, *Coimbatore*, *Nilgiri*, *Ramnad White, Madras Red* and *Tiruchi Black* are popular. Unlike breeds in the north, southern breeds of sheep do not have wool but are called hairy sheep. They are used mostly for dung and meat.

Forest based systems

In the heavily forested areas of the subcontinent, livestock has not played as important a role in the economy and livelihoods of people, as it has in the arid and semi-arid areas. The forest provides food, fodder, timber, fuel and medicines. Where forest resources abound, even agriculture is not very important. Livestock therefore, has not been developed for producing milk or wool but rather for adapting to forest conditions including the presence of wild predators like the tiger. Depending on the forest, buffalo, pig, elephant, the *mithun* and chicken have been domesticated. The wild progenitors of these species were themselves inhabitants of the forest. Although cattle and goat are also reared in forests they have been a later introduction. Unlike in the grasslands, livestock cannot play a major role within forests, as it would lead to impossible conflicts with other natural systems.

A large number of distinct tribal communities live in our forests and their livelihood is dependant on the survival of these forests. While there are communities who still mainly hunt and gather like the Katkaris of the Western Ghats, most tribal communities practise some agriculture. In many forests, slash and burn agriculture continues to be practised. The crops grown are mainly different millets, some pulses and a very few oil seeds.

Major Forest Based Livestock Systems in India

The eastern Himalayas

These hill ranges form the northeastern borders of India. As opposed to the northwest Himalayas, these ranges are covered with evergreen forests and characterized by a warmer and more humid climate. Stretching from Sikkim to Nagaland and across the Brahmaputra valley to the hills of Meghalaya and Mizoram, these hills are the home of a number of tribal communities. While paddy is grown in the valley, a variety of millets, maize and horticultural crops are also cultivated on the slopes. Here nomadic sheep and goat rearing gives way to a variety of other species from the elephant, yak and *mithun* to buffalo, goats, pigs and chicken. In the lowlands, duck rearing is popular. Unlike the vast open systems of the pastoral communities, animals here are reared close to the houses of farmers and very often stall-fed. Animals are mainly reared for meat and only occasionally for milk. In the Northern hills of Sikkim and Arunachal, the yak becomes an important species providing a means of transport, food and fibre. Yak dung is used as a source of fuel. Elephants are used in the more dense forest patches to help with the hauling of timber. The *mithun*, a unique animal descended from the wild gaur is also domesticated here. These animals have an important socio-economic and cultural significance amongst the people who rear them. They are sacrificed for meat and used as currency. Pigs are also a measure of wealth in these regions, and unlike the rest of India where pigs are mainly reared under open scavenging systems; pigs here are also well managed under closely monitored stall systems. The cattle breed found here is the *Siri*, which is suitable for draught work on the hill slopes.

Lower gangetic plain

This region comprising the lowland deltaic plains of the Ganges and Brahmaputra rivers is characterized by the marshy saline swamplands of the Sunderbans. The main crops are paddy and jute. Interestingly, this region is home to perhaps one of the very few breeds of sheep adapted to wetlands, the *Garole*. Reared mainly for meat this breed is known for its high prolificacy and twinning capabilities. The well known breed of goat from this region is the *Black Bengal*, small in size but with excellent quality meat and high prolificacy.

The central plateau and hills

Characterized by deciduous forest, this region spreads across the states of Orissa, Jharkhand (formerly South Bihar) Madhya Pradesh, Chattisgarh and parts of Eastern Maharashtra. Extremely rich in natural resources and minerals, this again is the home of a large number of tribal communities like the Kol, the Oraons, Gonds and Santhals who practise a mixed type of farming which includes a range of crop and livestock species. As this is rather heavily supplemented by forest produce, little or no specialization of agriculture or livestock takes place.

There are very few well known or distinct breed of animals from this region, which only reconfirms the fact that livestock does not play a major role here. Some of the tribes raise pigs. The *Sambhalpuri* and *Nagpuri* buffalo breeds that are reasonable milkers are found in this region. Breeds of goat like the *Chota Nagpuri* and the *Ganjam* and the *Ganjam* breed of sheep are also found here.

The Eastern Ghats

This region is almost contiguous with the region above and covers the lower parts of the state of Orissa and descends into the state of Andhra Pradesh. Dense semi-deciduous forests cover this region. The forests are central to the lives of the tribal communities who live here. Agriculture is based on some paddy, a variety of millets and pulses. Some of the interesting breeds of buffalo from this region are the *Manda*, *Jerangi* and the *Kalahandi*. Male buffaloes are often used for draught purposes.

The breeds of cattle found here are small and are kept mainly for draught and dung. In fact, these animals are seldom milked and the milk is essentially left for the calf.

The species that is the most diverse here are the poultry. There are many famous breeds of poultry from this region and people here identify over eight distinct strains of the *Aseel*.

Box - 2

The Poultry Breeds of India

Gallus gallus or poultry was reported to have been first domesticated in the jungles of the Indian subcontinent from its wild progenitor, the red jungle fowl. Since then poultry rearing has spread throughout the world. In India, besides eggs and meat, birds are also reared for cockfights and to be given away as bride price or dowry. This activity is almost entirely managed by women. The most famous breeds of poultry are found in the eastern states of the country.

Breeds of the Northeastern states

Chittagong

Also known as Malay these birds are found in the northeastern states of the country. They are large birds with the male birds weighing from

(contd.)

(Box - 2 contd.)

4 to 5 kgs and females from 3 to 4 kgs. These are good game birds being both strong and hardy.

Daothigir

These birds are found in Assam and reared by the Bodos. This is a fairly heavy breed with a good growth rate.

Frizzle fowl

The northeastern regions and the hot humid coastal regions of the country are home to this breed. Known for excellent adaptability in a tropical climate the rachis of the feather is curved due to the presence of a dominant frizzle gene due to which plumage looks curly. This feature increases the heat dissipation from the body .

Haringhata

A small-bodied black bird found in the state of West Bengal. This bird has the typical conformation of a layer bird.

Miri

This bird gets its name from the Miri or Mishing tribe who mainly rear this bird in the state of Assam in northeastern India. It is a small sized black bird.

Naked Neck

As the name indicates, the neck of the birds do not have any feathers or have only a tuft of feathers in the front of the neck above the crop. These birds are ideally suited for hot tropical climates. Birds of this breed are found in hot and humid coastal regions and the northeastern states.

Birds of Andhra Pradesh

The state of Andhra Pradesh has some of the finest breeds of poultry including the famous *Aseel.*

Aseel

The original home of these breeds is Andhra Pradesh. Well known for its stamina and majestic gait, this breed is popular with those who practise the sport of cock fighting. The standard weight varies from 3 – 5 kgs

(contd.)

(Box - 2 contd.)

for cocks and 2 - 4kgs for hens. These birds are also famous for their delicious meat.

Ghagus

Nomadic groups in the states of Andhra Pradesh and Karnataka rear this bird. This is a small- sized bird with small combs and wattles. The neck is thick and loose and the shank is feathered. These birds are hardy and supposed to be resistant to a number of common diseases but are today becoming scarce.

Denki

Also from Andhra Pradesh, the Denki is a fairly heavy bird with a glossy and lustrous plumage and a compressed single comb. Cocks are good fighters and the birds in general are fairly resistant to diseases.

Kalasthi

Similar to the Denki, but smaller in size these birds are popular in the state of Andhra Pradesh.

Birds of the west coast

Ankaleshwar

From Ankleshwar in the state of Gujarat, this breed is small and has a single comb. Birds are hardy and resistant to disease.

Busra

Found in the region around Bombay in the states of Maharashtra and Gujarat, these birds are small to medium-sized and light-feathered.

Tellicheri

These birds take their name from Tellicheri, a small town in North Kerala. These are small birds with black skin. The plumage varies in colour from black to grey and sometimes various combinations of colours. The meat of these birds is said to have medicinal value.

Birds from the Nicobar Islands

A native of the Nicobar Island, these birds are round, compact, small

(contd.)

(Box - 2 contd.)

and short with a stout neck. They are of a brownish matte colour, and are well adapted to thrive in tropical environments. These birds are the best layers amongst native breeds.

Birds from other parts of the country

Brown Desi

Abundantly found in U.P, these birds are of the layer type, light to deep brown in colour, and have a single comb.

Kadaknath

This unusual bird from Madhya Pradesh is also known as *Kalamasi*, meaning fowl with black flesh. The skin, beak, shank, toes and soles of feet of this bird are a dark grey. Even the internal organs show a black pigmentation. This dark pigmentation is due to the deposition of melanin. Despite the strange colouring, the flesh is supposed to be delicious. These birds are of medium size and are medium layers as compared to indigenous breeds that can lay up to 80 eggs per year. It is bred by tribal communities in the Jhabua and Dhar districts in Western Madhya Pradesh. Under free range conditions, these birds are resistant to disease.

Faverolla

This small bird with small comb and wattles is from Kashmir. A unique feature of this bird is its feathered combs.

Punjab brown

This bird is found in the states of Punjab and Haryana. It is a meat type. Males weigh 3 - 4 kgs and females weigh 2 - 2.5 kgs.

Tetri

These are small birds with speckled black and white feathers, yellow beak and legs. Birds are poor layers.

Teni

These are small-sized all purpose birds found all over the country and

(contd.)

(Box - 2 contd.)

are ideal for conditions in Indian villages. They tolerate high temperatures, typical of Indian summers, can forage actively for food. They are also good brooders i.e., they sit on eggs laid and make good mother hens.

The specialty of Indian breeds

Most Indian breeds are poor layers. Averaging from 45 – 80 eggs a year, they compare rather poorly when compared to breeds like the white leghorn that has the capacity to lay up to 200 eggs per year or more. On the other hand, the exotic varieties introduced into the country face problems of drop in egg production, low feed efficiency, thin shelled eggs and high mortality during the summer months due to heat stress. This is where Indian breeds score over exotics. They can tolerate heat, are resistant to disease and are supposed to be superior in taste.

Breeds of ducks

Ducks are reared in those parts of the country where plenty of water is available and are therefore in the eastern and southern states of the country such as Assam, West Bengal, Orissa, Andhra Pradesh, Tamil Nadu, and Kerala. They are frequently raised in paddy fields. The known Indian breeds are *Indian Runner*, *White Bearded*, *Nageshwari* and *Sylhet.*

Breeds of guinea fowl

The Guinea fowl is also reared in many parts of the country. There are no defined breeds but pearl, lavender and white varieties have been documented.

Pigeons

Pigeons are also popular in parts of the country. In fact till recently the Orissa State government had a pigeon post.This service has been now discontinued.

Source: Sahai, R. and Vijh, R.K. (Eds.). 2000. *Domestic Animal Diversity Conservation and Sustainable Development*. Karnal: SI Publications.

The Western Ghats

These hill ranges start to the north of the state of Maharashtra and stretch all the way down to Kerala running parallel to the west coast of the country. The main feature of this mountain range is heavy rainfall. Earlier these ranges were covered by very dense evergreen forests, which were almost impossible to penetrate. Where the forests were dense, the buffalo was the preferred species as it could withstand heavy rain, thick forests and confront wild predators. The Gawli Dhangars, a community of traditional buffalo rearers live in these hills in the region between Maharashtra and Karnataka and their livelihood depends on the sale of milk and milk products like *ghee*, buttermilk, yoghurt, *khoya* and milk sweets like *pedas*. Today with loss in forest cover these communities rear cattle as well. Further south in the Nilgiris, the *Toda* is a breed of buffalo, which is reared exclusively by the Toda tribe.

The *Dangi* is a special breed of cattle found in the Western Ghats of Maharashtra. This breed is supposed to be able to withstand heavy rain due to the presence of a special oily secretion by its skin. It is reared largely by the Mahdeo koli and Thakar communities. The *Dangi* is the preferred cattle breed of this region as it is able to withstand the climate characterized by extremely heavy rainfall, work in the hilly terrain as well as yield reasonable quantities of milk provided there are enough forests to graze in. For these communities the female *Dangi* is 'Lakshmi', the Indian Goddess of wealth and prosperity. Another identified cattle breed found in the Western Ghats is the almost dwarf like *Malnad gidda* of the Malnads or uplands in Karnataka. The reason for this short size has different explanations. For one, the small size makes it easy for the breed to plough in the small pocket-sized landholdings, which are cut into the hill slopes. Secondly, the soil is lateritic and acidic and the crops and grasses here are deficient in calcium resulting in a small bone structure and skeletal growth. Thirdly, inexpensive animals means the loss to the owner is relatively small if one were to be taken away by a wild animal or bitten by a snake, which happens often enough.

In the forests of Karnataka and Kerala, elephants are domesticated for helping with hauling timber. Many farmers also keep goats in this region because of their adaptability to the hill terrain and the fact that they are easy to rear. The *Malabari* from Kerala is a breed from this region.

An interesting feature is farmers from the adjoining dry Deccan plateau bring their sheep to the wet uplands to graze during the winter and the summer. Like with the other forest areas, poultry are popular amongst farmers in this region. Breeds like the *Naked Neck* in the Sahyadri ranges and the *Busra* are found here. As with most Indian birds these are meat birds and are not very good layers.

The Coastal Regions

India is a peninsula and has a rather long coastline. People living on the coast have livelihood patterns unique to their region. Just as the forest provides so does the sea. Most people living on the coast consume seafood and therefore do not necessarily need other animal protein. However, seafood is not available all the time and many coastal communities also practise agriculture and livestock rearing which is however quite distinct from pastoral and forest based systems.

Here again there are differences between the eastern coast and the western coast of the country both in geographical and ecological terms. Consequently, practices vary quite substantially between the two coasts.

The west coast

The west coast of the country is uniformly characterized by heavy rainfall and lateritic soils. The main crop is paddy. While the size of the plots for paddy on the coast in the state of Maharashtra, are small they increase substantially as one moves south into Goa. Running parallel to the coast are the hill regions of the Western Ghats. Where the hills begin, food crops like millets give way to cash crops like rubber, mango, spices and cashew nut for which these regions are famous.

Only those animals which can survive the heavy rainfall are found here. The soils here are acidic and the cattle found here is very small as typified by the *Vechur* of Kerala, a cattle breed which is no larger than a goat.

Breeds of goat are found all along the coast. The *Surti*, a breed of goat perhaps derived from small Arabian goats which were brought by Arabian traders are found in the coastal district of Surat in Gujarat. They are very good milkers. In Kerala, the *Tellicherry* breed is found in and around the coastal town of Tellicherry. Pigs are reared all over the coast but are more popular in Goa.

The east coast

The east coast is quite unlike the west coast. It is broader and the plains are formed by the deltas of many large rivers, the Mahanadi, Godavari, Krishna and Cauvery. This region is also prone to severe cyclones. These plains are extremely fertile and paddy and coconut are the principal crops. One of the best known Indian breeds of cattle, the *Ongole*, is the favourite draught animal of the coastal plains of Andhra Pradesh. This breed renowned in India for its draught and dairy capabilities is bred in other countries as a meat animal.

There are breeds of sheep and goat found here as well which are reared by communities like the Gollas. These shepherds migrate with their animals from the coast upwards into the hills of the eastern ghats and back again. Sheep breeds found here do not have wool, but are covered with hair and are mainly kept for meat and dung. Both the sheep and goat breeds go by the name of *Nellore*.

The Islands

The southern islands of Andaman and Nicobar and the Laccadives have an economy primarily based on forests and coconut and the sea. Pigs are the preferred species reared by communities like the Jarawa, Sentinilese and the Onges. The breed of pig found here is quite distinct from the pig found in the mainland. These animals have an enormous economic as well as socio-cultural significance in the lives of the people who live here. Another favoured species is the goat. The *Nicobari* chicken enjoys the reputation of being the best layer amongst the chicken in the country.

Crop Based Systems

As mentioned earlier, the big revolution in civilization probably occurred when crop farming and animal farming were brought together. This enabled the creation of surpluses, which could then be traded for money. In India, the yoke fell on the male of the species *Bos indicus* otherwise also known as *zebu* cattle were used to assist in agricultural operations. The presence of a distinct hump on the back of these cattle made it possible to yoke the plough easily. It was also possible to attach a cart to a pair of oxen, which could then be used for transport.

Indian farmers, therefore, began to carefully select and breed cattle more for their draught capabilities than for milk. Size, stamina, resistance to disease, temperament, size of hooves besides the colour, shape of horns and position of whorls that played an important role in determining the most suitable breed for a particular region. The breeds in the mountain areas, for example, had to be small and light while the breeds of the plains needed to be heavy. Areas with clayey soils needed animals with large hooves while areas with light lateritic soils needed animals with light hooves.

In certain areas where it was extremely difficult for cattle to work as in the wet paddy fields of coastal areas, male buffaloes replaced cattle as work animals. In extremely dry desert areas the camel made for more efficient use than cattle, but by and large throughout the subcontinent cattle began to gain importance as a draught animal and this practice continues till today.

For centuries, cattle have ploughed the fields and helped farmers cultivate their fields. The draught breeds of India are magnificent animals and in their honour special festivals like *Mattu Pongal* in Tamil Nadu and *Bail Pola* in Maharashtra are still celebrated.

A common criticism of our farming systems is the large number of male animals which are put to use during a very small period of the year and which lie idle for a greater part of the year. However, there appear to be several reasons for this as vouchsafed by small farmers. There is a very small window in the agricultural calendar, which is available to farmers for ploughing. This window is dependent on the monsoons, which are best unpredictable. During this short period of time all farmers need to complete preparing their field for sowing. Failure to do this on time means no agricultural crop and consequently no food the following year. Rather than face that eventuality, farmers would rather maintain a pair of bullocks so that they can carry on with their agricultural operations as per schedule rather than be forced to borrow or hire a neighbour's pair of bullocks. Cattle could also be used for threshing the harvested crop. The seemingly 'useless' bovine also provided transport for hauling farm produce over long distances and valuable dung which was both fertilizer and fuel in a pre-petroleum era. The bullock, therefore, had multiple utilities and even today is indispensable to the existence of poor farmers.

The number of bullocks or oxen as they are also known is often a direct reflection of the agricultural land owned. The more the land the greater the number of animals one needed for work as well as for manure. To keep a large number of animals one also needed enough fodder, which was available as crop residue from the agricultural fields. Dairy animals were a luxury only permitted to wealthy farmers who had surpluses of feed. These households converted surplus milk into ghee for household consumption and ritual offering, while the by-product—buttermilk—was given to servants or farm labourers.

Otherwise dairy cows were mainly kept by specialized communities like the Yadavs of North India, the Ahirs of North and Central India and the Patels of Gujarat who practised some crop farming but who also specifically reared dairy animals and earned a livelihood selling milk.

Small and marginal farmers who were forced to depend on common lands like grazing lands and forests often settled for breeds and varieties, which did not require much maintenance and consumed smaller quantities of fodder. Landless farmers normally kept goats or poultry or sometimes did not keep animals at all.

To say that agriculture in ancient India was not productive would be grossly incorrect. To quote Sir Albert Howard from his *An Agricultural Testament* (1940):

> 'In the agriculture of Asia we find ourselves confronted with a system of peasant farming which in essentials soon became stabilized. What is happening today in the small fields of India and China took place many centuries ago. There is no need here to study historical records or to pay a visit to the remains of the megalithic farming of the Andes. The agricultural practices of the Orient have passed the supreme test –they are almost as permanent as those of the primeval forest the prairie, the ocean.'

In *Indian Science and Technology in the Eighteenth Century: Some Contemporary European Accounts*, Dharampal(1983) has further demonstrated the close relation between livestock and crop rearing in India:

> 'A plant called sota gowar is sown broadcast with sugarcane in Guzerat. The gowar serves as a shelter to the sugar cane from the violent heat of the sun, during the most scorching season of the year. Joar and badgery are sown together, in the same country late not for the sake of a crop, but for straw, which is very nutritive and abundant. This is one of the instances in which the native provides a green crop for their cattle. Other grains are sown both together and separately merely for their straw. Soondea, darrya joar, rateeja and googra joar are sown together but with the exception of googra joar which is allowed to ripen, the rest are reaped when still green.'
>
> It is evident that these examples are not founded on bad principles, but that they lie in conformity with the best practice of farming. They evince the care of the Hindoo Husbandman to provide food for his labouring cattle. This is an object to which I have generally seen him very attentive: but in many parts of India during the dry season it is extremely difficult and often exceeds the impoverished means of the cultivator, to lay in a sufficient supply. He is sensible enough of the want and does his utmost to scrape together, all the heterogeneous substances that are within his reach. In some parts of India, hay is not made, in other parts it is a regular crop, stacked and preserved.

'In those parts of India where hay is not made and which are I believe unfavorable to the crop, the cattle are fed with the roots of grass, very like our fiorin, with straw and especially with the straw of joaree all of which are considered to be very nourishing food.'

The author further states that the ...'the whole world does not produce finer and more beautifully cultivated fields than those in Guzerat. He also goes on to suggest a little later that, ...this appearance is not peculiar to Guzerat: It may be found in many other countries of India'.

The above excerpts are not reproduced to extol traditional Indian practices but to draw attention to the fact that the present-day criticism of Indian agricultural practices is often faulty and misplaced and the close relationship between the draught animals of the country and agriculture should not be ignored or forgotten.

Let us take a closer look at our draught animals and where they are found.

The purely draught breeds of cattle are:

1. ***Nagori:*** This breed is found in the Nagaur, Ajmer and Jodhpur districts of Rajasthan. Animals are useful for heavy draught work and transportation in deep sands. Today, this breed is getting displaced by the tractor.
2. ***Bachaur*:*** This breed is from the Sitamarhi district of Bihar and can work under stressful conditions of this region.
3. ***Kenkatha*:*** Animals of this kind are found in the Bundelkhand region of Uttar Pradesh and Madhya Pradesh. The name is derived from the Ken River that flows in these regions. They are small, sturdy and powerful animals.
4. ***Malvi:*** This breed is found in the western regions of Madhya Pradesh and some parts of adjoining Rajasthan. These animals are good for working on black cotton soils.

---

*Threatened breeds

5. ***Kherigarh****: This is a breed from the Lakhimpur Kheri districts of U.P. They are small in size but very active and are good draught animals suited to the *Terai* tracts of the country.

6. ***Hallikar***: Found in the Mysore region of Karnataka, this animal is alert, strong and nimble footed. Males are supposed to run like deer, as they are extremely light.

7. ***Amrit Mahal***: This is the draught breed found in the Hassan, Chikmaglur and Chitradurga districts of the state of Karnataka. The breed was specially developed by the Maharajas of Mysore. They earned their name in the reign of Tipu Sultan when they were put to use for pulling heavy guns during war.

8. ***Khillari:*** These animals are found in the Kolhapur, Solapur, Sangli and Satara districts of Maharashtra and the adjoining districts of Karnataka. Aggressive and strong, this white animal with sharp horns pointing backwards is the dream of every farmer in the region. Many traditional songs are dedicated to this breed. There are *Ovis* (songs sung by women while grinding grain in the mill stone) that describe in loving detail how a *Khillari* bullock must be. The favourite names given by farmers for these animals are Raja (king) or Nandi the animal of Lord Shiva.

9. ***Bargur***: Found in the Erode district of Tamil Nadu, this breed is suitable for agricultural operations in hilly terrains.

10. ***Kangeyam***: Like the Khillari of Maharashtra, this is the famous breed of Tamil Nadu. It is found mainly in districts of Coimbatore, Dindigul, Erode and Namakkal. These animals are docile, quick and strong, and thrive on scanty rations. They are bred by big cattle breeders of the area known as the Pattegars (Randhawa).

11. ***Umbalacherri***: This breed from the fertile soils of Tanjore district of Tamil Nadu has a distinctive bald appearance as its horns are burnt out when young. Smaller than the *Kangeyam*, it is good for ploughing paddy fields in this region where puddling is practised. Bullocks of a heavier breed would run the risk of sinking too deep in the soil.

12. ***Ponwar*:** This is a small hill type cattle found in the foothills of Pilbhit region of Uttar Pradesh.

13. ***Siri**:** This is a small breed found in the hilly terrain of Sikkim, West Bengal and the neighbouring countries of Bhutan and Nepal. It is especially good for operations on hill slopes. A special thick skin with supposed water resistant qualities makes this breed suitable for high rainfall areas.

Besides these draught breeds, which are only suitable for agricultural work, there are many multipurpose breeds also known as general utility breeds, which are reared for both draught and milk.

Some of the well-known breeds in this category are:

***Nimari*:** This is a breed from the Nimar tract of the Narmada valley of Madhya Pradesh and adjoining Maharashtra.

***Hariana*:** This is a famous breed found in the Rohtak, Hisar and Gurgaon districts of Haryana state. Recognizable by the prominent poll between the horns, this breed is white in colour, sleek and rather attractive in appearance.

***Mewati*:** Similar to the *Hariana*, this breed is found in the Mewat region of Rajasthan as well as the adjoining parts of Haryana and Uttar Pradesh.

Kankrej: This is a particularly majestic breed, large and heavy and extremely popular in the region of Rann of Kutch, the adjoining districts of Rajasthan in India where it is called the *Senchore*, and also in Sindh in Pakistan. These animals have horns that are large and curve inwards. Animals of this breed have a particularly attractive slow swinging gait. Maldhari, Bharwars and Rabari herders maintain large herds of these cattle. This breed can be seen all over Gujarat and western Rajasthan even today.

Gaolao: Found in the Vidarbha region of Maharashtra, this breed is supposed to have been developed by Maratha warriors for transportation purposes in the hilly areas of the central plateau.

Red Kandhari: These animals are found in the Marathwada region of Maharashtra especially in the districts of Nanded, Latur and Parbhani. They are red in colour and heavily built. These bullocks are particularly suited for heavy work in the black cotton soils of the Deccan.

Deoni: An excellent dual-purpose breed that gets its name from a village of the same name in the Latur district of Maharashtra. The Deoni is extremely popular amongst farmers in the region as well as in the adjoining districts of the neighbouring states of Andhra Pradesh and Karnataka. The females are fairly good milkers under well managed conditions, while the males are well known for their mild disposition which makes them easy to handle.

***Krishna valley*:** This breed is supposed to have been developed in the last part of the nineteenth century by combining the *Gir*, *Ongole* and *Kankrej* with local animals. This breed is particularly good for agriculture work in heavy black cotton soils found in the tract. The districts adjoin those of the Khillar such as Solapur, Sangli, and Satara districts of Maharashtra and includes some adjoining districts of Karnataka and Andhra Pradesh.

***The Ponganur**:** Supposed to be one of Asia's shortest cattle breeds, this breed is found in the Chittoor district of Andhra Pradesh. Although the male is sometimes used for draught the females are fairly good milkers needing extremely small quantities of fodder to sustain them. The breed is reputed to have an excellent feed conversion ratio. There are less than a hundred of these animals today which puts these animals in the highly endangered category.

It is not just male oxen, which are important for agriculture. Besides cattle, a number of other species of animals also assist in agriculture. In the more wet and marshy areas male buffaloes often take the place of cattle. The draught breeds of buffalo are mainly found in the coastal and hilly districts of the states of Orissa, Andhra Pradesh and Kerala. The documented breeds are the *Sambalpuri* of Orissa and the *Manda* and *Jerangi* of Andhra Pradesh and Orissa. In the arid areas and semi-arid areas, camels are put to work in the agricultural fields. Popular draught camel breeds are the *Bikaneri* of Bikaner, *Marwari* of Jodhpur, Jalore and Barmer regions of Rajasthan and the *Mewati* of Alwar and Bharatpur in Rajasthan.

While cattle may be the best known for their contribution to agriculture, many other species also contribute substantially. Sheep and goat play an equally important role in agricultural operations in India. As mentioned before, farmers in many parts of the country invite shepherds to fold their animals in their fields. Of all the types of manure available, sheep manure is considered the best. In the northeastern parts of the country where pigs are raised pig manure is used to fertilize fields and the wastes from the kitchen and fields are used in turn to feed pigs.

Similarly, in many of the Eastern states of the country ducks are reared in rice fields. The ducks obtain their nutrition from the fields in the forms of weeds and insects and in turn keep the fields free of unwanted weeds and pests.

Different synergies have thus been developed between different breeds, farming systems and different communities with each of them providing a distinct and unique pattern. The study of traditional systems provides an insight into these different patterns and the reason for the differences.

Species of Domination: The Ruling Class of Animals

Finally, over and above agriculture, the history of domestication progressed in another direction, that of domination.

The draught capabilities of animals were essential in times of war and kings and rulers began to develop species and breeds to suit certain special requirements. Horses were specifically bred for riding and so were camels. Elephants became symbols of power. Cattle were needed to provide milk and nourishment for the army as well as for transport. In fact, the game of chess, which has originated in India clearly gives the position of different animals in the royal courts.

The link to royalty added a totally new dimension to the care of animals. The science of animal rearing, management and health care grew highly developed and was codified into texts and rituals. There were details on the appearance of different species, auspicious characteristics on their body by which they ought to be selected and inauspicious characters by which they ought to be rejected. The best diets were prescribed and so were treatments for various conditions. Different communities of animal rearers were responsible for breeding animals and supplying rulers with the animals they required.

Animals were also important for trade as they could transport goods over long distances. Caravans of traders traversed the hills and plains of the Asian continent establishing trade routes and exchanging practices. New strains and breeds of animals which entered the country crossed with local breeds. Those that could adapt survived, while those that could not adapt naturally died. This gave rise to new and interesting breeds of animals, especially amongst horses and camels.

Animals also went as gifts and dowry from one kingdom to another. With them went their keepers and their trainers who then passed on their traditions to new areas.

Extensive treatises and texts were written by experts for the benefit of the royal court and its members. These texts minutely detailed how these species were to be taken care of and managed.

Some of the earliest works were the *Krishi Purana* and the *Agni Purana*, which give details on how animals were to be selected, managed and cared for. The *Rig Veda* details responsibilities of herders towards their animals. The *Krisi Parasar* is another treatise that contains elaborate rules on the management of cattle.

The *Mahabharata* of course tells us stories of Krishna, the Yadava king, his fondness for butter and other milk products. But besides that, it also talks about Nakula and Sahadev, the twin Pandava princes who enjoyed the reputation of being experts in the care of horses and cattle.

Kautilya's *Arthashastra* carefully details duties of livestock herders, veterinarians and the penalties to be imposed on them for failure to do their duty. It also gives details on the management of fodder and pasturelands, of forests and their use.

Later texts include the *Gaja Shastra Sara* of Rakhmaji Pandit which is a summary of the great work on elephants by Sage Palakpya, *Hasta Ayurveda*. There are also the *Asva Shastra* of Hemasuri, a summary of the work by Sage Shalihotra and the *Brihat Samhita* by Varahamira which deal with the care of animals.

There are ancient palm leaf manuscripts in Tamil called *Pasu Vaidya Mattu Vagadam*, which deal with two hundred and fifty diseases and other conditions affecting cattle. Much later in the reign of Raja Serfoji, the Saraswati Mahal library of Tanjore brought out many books based on manuscripts found in Tamil, Marathi and Sanskrit on different subjects of livestock care.

This codified knowledge however, was essentially available to those who were close to royalty or worked in the palace farms and stables or for those who could read and write. This largely left out the peasantry who developed their own systems based on empirical observations and passed them on as oral traditions from one generation to the next generation.

Over the years, rulers and dynasties changed. Each change brought with them new sets of practices and demands to suit their needs. However, many of the practices known to farmers, pastoralists and forest people who were not close to the seats of power remained scattered and diffused. Yet they constitute a valuable set of practices and is what classified as indigenous knowledge or ethno-veterinary practices.

3

Traditional Practices in Animal Rearing

The diversity in livestock species and breeds and the people who rear them in India is a reflection of the diversity in ecology. The different systems are also a reflection of the forces that influenced the type of system extant. Traditional practices of housing, breeding, nutrition and management were therefore greatly influenced by the agro-ecological diversity of the region, the patterns of rearing and the people who reared them.

Traditional Breeding Practices

Given the large diversity of breeds in the country, one is left in no doubt that careful breeding has gone on for centuries to mould and shape a particular species from its wild state and produce a domesticated species, which yields animal products that can satisfy different needs. There has also been a process of natural selection whereby the strongest animals, or rather in other words, those most resistant to a disease or animals most suitable to a particular environment have survived as breeding stock.

Breeding has always been a selective activity and keeping a quality animal is an expensive proposition. The best breeds were always found belonging to the local rulers. Indian breeds of cattle have attracted international attention from as far back as the nineteenth century when breeds like the *Gir*, *Ongole* and *Sahiwal* were exported to the Americas and Australia under the name Brahman bulls. This was the time when Europe was busy colonizing. The grasslands of the Americas, the prairies and the pampas as well as Australia and New Zealand offered vast opportunities for large-scale cattle farming. Excellent animals were taken to these countries by the colonizers. Careful breeding and selection have also taken place outside the palace portals.

Even in animals that exist in a semi-domesticated state like the *mithun* and the yak, a certain degree of selection has gone towards evolving suitable animals. In the case of the yak, bull calves showing larger body size and having attractive colours and horn patterns are selected as sires for breeding (sire is the technical term for the male parent). In some areas, yak and *Siri* cattle hybrids have also been produced for areas, which are too low in altitude for the yak and too high in altitude for cattle. The Monpas, a Buddhist tribe of the West Kameng and Tawang areas of Arunachal Pradesh have used *mithuns* for hybridisation with the local *Siri* cattle.

Selection for animals has always been in response to the production goals of the community which is raising these animals. Horses and camels in India have been developed both as riding animals and as pack animals.

Breeds which are used for riding are of lighter build and slender while those selected as pack animals are often of heavier build and more sturdy in structure. Cattle and buffaloes bred for draught do not have dairy qualities, but their selection is based on their ability to perform different agricultural operations. Animals that migrate are selected on the basis of their ability to withstand the pressures of migration and the stresses imposed therein. This is true of many of the breeds of sheep found in the semi-arid region of country. Breeds selected in areas with heavy rainfall have a thick skin that can withstand heavy rains like the *Dangi* and *Siri* breeds of cattle. Resistance to disease and parasites are some of the other factors which Indian farmers need to consider while selecting breeds. Amongst chicken, selection has largely been towards birds, which produce quality meat, can withstand disease, make good brooders and are chosen especially for cock fighting abilities, which also happens to be a popular sport.

Breeding strategies are very advanced in animals belonging to pastoral communities whose livelihood essentially depends on having quality stock. Although difficult to comprehend because careful pedigree records are not maintained there is a great deal of thought which goes into the process.

Male animals are carefully selected as breeding stock and retained in the herd while the rest are sold. Care is also taken to prevent inbreeding by changing the stock periodically or bringing in animals from other herds.

Box-3

Indo-Swiss Goat Development and Fodder Production Project (ISGP)

The Indo Swiss Goat Development and Fodder Production Project (ISGP) was established in 1981, through an agreement between the Governments of India and Switzerland. ISGP was implemented through the Rajasthan State Department of Animal Husbandry with technical cooperation from Intercooperation Berne.

The main objectives of the project were to develop strategies for sustainable improvement of goat production in the semi-arid farming systems of Rajasthan in order to improve the income-generating capacity and nutrition of families belonging to the weaker sections of the rural community.

The project was head quartered at Ajmer with a base farm at Ramsar 40 kms south east of Ajmer. Extension activities of the project concentrated on the indigenous *Sirohi* breed that has its habitat in the Aravalli hill range. In carrying out its work, ISGP placed a major emphasis on better understanding of the traditional practices and the dynamics of goat production in Rajasthan.

In order to assess the merit of cross-breeding for improvement of goat milk production in the semi-arid zone of Rajasthan, a comparison was made between the performance of the local *Sirohi* breed and its 50% crosses with Alpine and Toggenburg bucks imported from Switzerland.

While the crosses had a lower age at kidding, it was found that animals kidding at a younger age tend to have lower milk production because part of the energy is allocated by the body to growth. The *Sirohis*, however, came into heat earlier during the season, which meant additional days of lactation. The *Sirohi* goats also showed a lower rate of mortality in particular amongst kids as compared to the crosses. Based on these findings, it was concluded that it would not be economically justified to initiate a large-scale crossbreeding program for *Sirohi* goats in Rajasthan. Contrary to the existing earlier records, records collected by ISGP Sirohi goats showed an average lactation yield of 240 kg over a 180 day period with 25-30 % of the does (female goats) producing over 300kg which compared very favourably with the production capabilities of the exotics.

Source: Indo Swiss Goat Project, Ajmer. (1993). *Manual for Improved Goat production.*(Parts 1-4). Berne: Interco-operation.

Some of the crucial factors influencing selection are:

- Criteria related to or assumed to be related to production traits such as rate of growth of young ones, milk yield, etc.
- Known genetic defects of the individual animal, its parents or offspring. Such animals are rejected.
- Economic and logistic constraints :
  animals which demand heavy feed resources may not be welcome in resource-poor areas.

Breeding has also been a fairly specialized activity and there are a number of communities throughout the country whose livelihood depends on the sale of breeding stock. They charge other farmers for the services rendered by their animals. Often these are different pastoral communities like the Rebaris and Bhadawars. Owners of female animals are charged for a service and if the animal does not conceive it can be taken back for providing services two more times. This system exists for buffaloes as well. These services, however, were too expensive for poorer communities and they have had to be content with nondescript animals or animals which have been rejected by wealthier farmers.

In many parts of the country though, breeding animals are maintained as a community resource and the entire village treats these animals with a great deal of reverence. Buffalo and cattle bulls of good breeds are allowed to wander at will through different villages. They are also given extra nourishment and care and allowed to stay wherever they find it convenient. Bulls are often seen occupying the village panchayat house or the police station in Rajasthan during the day and it becomes almost impossible to make them move leading to considerable inconvenience in the pursuit of daily official duties. However, since these animals are revered they cannot be disturbed. In fact, in Rajasthan there is another practice of gifting a bull to the village in the name of an ancestor. The practice is called *Amar Saand*. Earlier, the practice was restricted to the higher castes but has later been extended to other castes as well. This practice is partly the reason why there is considerable opposition to castration programmes in these areas.

Animal Housing

Housing protects animals and provides them shelter from inclement weather and predators. In the case of animals belonging to royalty, elaborate stables and sheds were built within the palace premises. Wealthy farmers also tried to provide comfortable houses for their prized animals. One has even heard of prized cows in Marathwada being housed in the main living room with a fan and cooler specially installed for the animal. By and large though, shelters for animals have depended largely on the biogeographic region, which in turn influences the natural resources available. Most often the houses are next to the farmers', own residence. Sometimes animals and humans share the same space. The material used for building the shelter is from locally available stones, bricks, wood, even crop residue and leaves depending on what the farmer can afford.

Under the pastoral system, except in the very cold regions of the Himalayas, animals were usually raised in the open. Large flocks of sheep were penned in fields sometimes within a net but often with just a hedge of thorn or bramble to keep them from going out and to prevent other animals from coming in. As the rainfall in these areas is very low, there is little danger of animals getting wet. The primary protection is from the cold in winter, when animals huddle together. Inevitably, the herdsman also spends his time close to his animals if possible under a tent.

In the heavy rainfall areas, and in very cold areas, animals have to be protected from rain, especially in the case of small ruminants and poultry. Shelters of animals in these areas are often under the house of the farmer. Since there is also a danger from wild predators, these animals are usually kept close to human habitation. Often, small animals are penned in sheds with slatted floors, to prevent the accumulation of waste within the pen.

Farmers also invest different materials into their housing. In the coastal areas of the east, which is cyclone-prone, houses are usually circular with low roofs and of light material.

Houses for poultry also take different forms. Poultry are most often, reared in baskets of bamboo, as they are easy to transport in this way, but interesting forms such as little woven shelters hung on trees with woven ladders which are pulled up at night are also seen.

Some of the most interesting types of housing for cattle are found in the western coastal region of Ratnagiri. This area experiences extremely heavy rainfall in the monsoon. Being warm and humid, the area is also ideal for the proliferation of ticks. Farmers build beautiful shelters that have walls made of *chira*, the local name for lateritic blocks. The roof is made of tiles and has a loft for storing paddy straw. Sides that are open to the rain have leaves and bamboo splints holding them together.

Each house is expensive to construct, but farmers realize it is crucial that the animals are healthy and well during the agricultural season which coincides with the rains. After the rains, the house is dismantled block by block. All the bamboo and leaves are burnt. This way any ticks nesting in the area are destroyed. Animals are then tied for a few days in the open some distance away from the main house. This is to ensure that the life cycle of the ticks gets terminated. A fresh shelter is then constructed with four bamboo polls as a frame and an interwoven mesh of leaves and twigs acts on the walls and roof. This then provides shelter for the animals in the mild winter and humid summer season.

Interestingly enough, during the earthquake of 1993, which had its epicenter at Latur in the state of Maharashtra, it was found that fewer cattle died as compared to poultry, sheep and goats. While poultry, sheep and goat were kept within the main house of the farmer which was made of stone blocks, cattle were housed in simple shelters made of the stalks of pigeon pea or woven grass and twigs so that, on the event of a collapse, the animals are not hurt (Ghotge,1995).

A study of the different housing practices can themselves form a book. The focus, however, in this section is on illustrating the management practises which have evolved over a number of years and are still in practise in large parts of the country.

Nutrition for animals

Animals instinctively know the best food to eat. Given a choice, in the wild they are quite capable of selecting the best pasturelands for themselves. However, when feeding them becomes the responsibility of human beings, questions regarding diet, daily rations and the best food for animals need to be examined.

Pasture and grasses are undoubtedly the best nutrition material for animals. If animals have sufficient green grass there is almost no need to feed them anything else. However, in the tropics grass dries up in the summer and in the temperate regions of the country grass is not available in the winter and animals then have to depend on other resources or migrate. Depending on the type of grass available the country can be divided into different types of grasslands.

Figure 3.1 Grass Cover of India

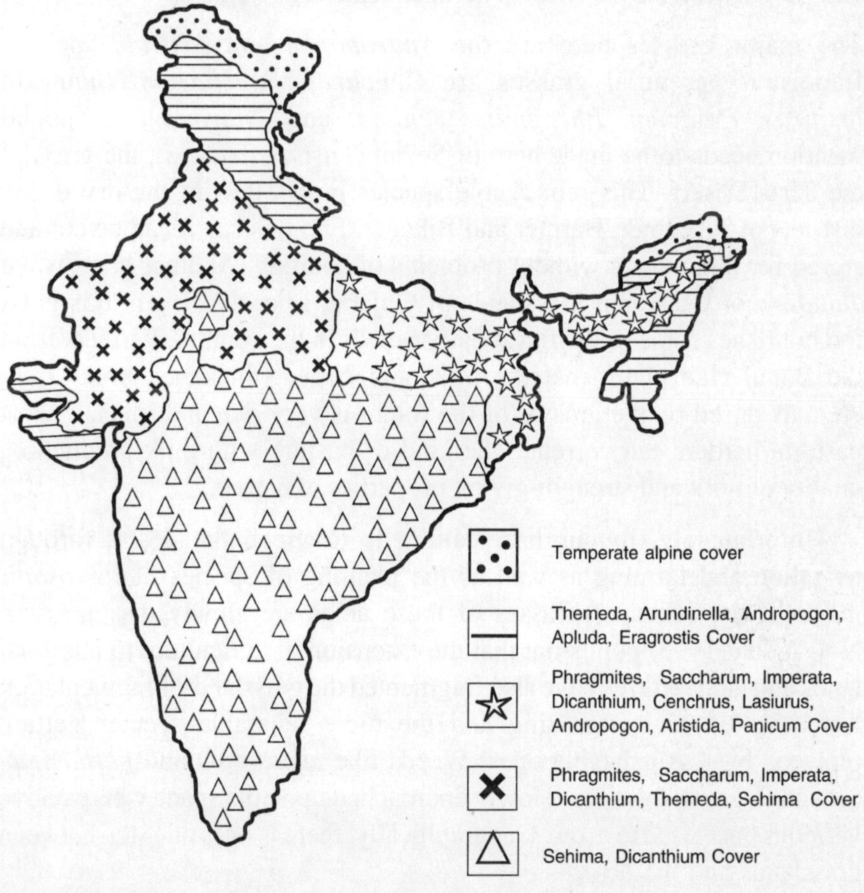

Grasslands of the Country

The northern temperate region

This region is characterized by temperate species which are available in summer in the upper ranges and tropical species that are available in monsoon and winter on the lower ranges. The northern pastures of the Himalayas have rich alpine legumes and grasses like red and white clover and rye grass. In the lower Himalayan ranges, the annual grasses found are *Themeda*, *Arundinella*, *Andropogon*, *Apluda*, *Aristida*, and *Setaria* and perennial grasses are *Eragrostis*, *Heteropogon* and *Themeda* in the Alpine pasture.

The northwestern, central arid and semi-arid regions

The major grasses here are the *Andropogon* and *Aristida* species. Important perennial grasses are *Cenchrus*, *Saccharum*, *Imperata lasiurus*, *Panicum*, *Phragmites sehima*, and *Andropogon*. Special mention needs to be made here of Sevan (*Lasiuris indicus*), the grass of the Thar Desert. This remarkable species grows best in the dry desert districts of Jaisalmer, Barmer and Bikaner. Farmers say it can be cut and stored for three years without problems of spoiling. Another grass is the *Jhinjhwa* of the Banni grasslands of Gujarat. Like the Sevan, this grass too nourishes some of the finest dairy breeds in the country. Farmers from the Banni clan claim that the milk and *khoya* (thickened milk) from animals raised on the grasses of the Banni is very rich and special. Most pastoral herders can correlate the kind of pasture with milk production, quality of milk and strength-giving properties of a grass.

Unfortunately, in our hasty attempts to green the desert through irrigation and farming as well as the planting of species like *Prosopis juliflora*, the principal grasses of these areas are slowly disappearing. N.S. Jodha (1995) points out that the extension of agriculture to marginal lands and land reforms have also fragmented the grasslands. Fragmentation has now led to overgrazing and the more palatable species getting replaced by less palatable ones. Weeds like *eupatorum* and *parthenium* (congress grass) have also slowly encroached upon the space where native varieties once used to grow. Geographically, there is only one step between grasslands and the desert.

Grasslands are ecologically sensitive and fragile areas, which have to be managed with great care. Left undisturbed, they can regenerate themselves to a reasonable extent but one small mistake can lead to disastrous consequences such as the creation of waste and barren lands, soil erosion and ultimately deserts. To prevent further desertification, it becomes critical to maintain grasslands. Grasslands are also important as reserves for important wild species, birds, reptiles and mammals. By destroying the of grasslands one may run the risk of driving a few more wild species towards extinction.

The southern region

The major annual grasses here are species such as *Aristida* and the perennial grasses are *Dicanthium*, *Sehima*, *Cenchrus*, *Chrysopogon* and *Themeda*. Unfortunately, here too due to overgrazing and bad management practices, the more palatable species like *Sehima* and *Dicanthium* have been replaced by the less palatable spear grass (*Heteropogon*), which although nutritively good has awns (spear-shaped seeds) in the winter months which hurt the insides of the mouths of grazing animals.

The eastern region

This comprises the eastern states of Orissa upwards into Assam and the northeastern states. The major species here are the *Andropogon*, *Apluda* and *Aristida* and a cover of *phragmites*, *Saccharum* and *Imperata*, The perennial grasses are *Dicanthium*, *Sehima* and *Themeda*. As the eastern corner of the country is by and large more humid than the west, the forest cover is more extensive and pronounced here and the grasslands do not assume any great significance.

Although grasslands can support all types of herbivores and ruminants, they are most suitable for sheep as sheep are grazers and survive best on land which has little tree cover. The grasslands in the country are also important in that they are home to a number of wild species for which India is so famous. Deer, antelope, and their predators, the big wild cats as well as a number of species of birds, reptiles and insects are found in these grasslands.

Over the years, these grasslands have got encroached upon and appropriated for diverse uses. As the demands of the human race have increased, the size of these grasslands has shrunk and due to overgrazing the more palatable grass species have been slowly replaced by less palatable ones. This has led to changes in patterns of livestock rearing.

Village Grazing Lands

Besides the grasslands where agriculture was not possible, villages in India also had demarcated grazing lands known by different names in different states. *Gochar*, and *gairan* are some of the well known names for these lands. Grasslands and their management have been mentioned in the *Arthashastra*, which was written about 2400 years ago. There were strict rules on their use and penalties and fines were imposed on defaulters. Later kings also paid attention to this critical issue because the wealth of their land and their animals depended on these pasture lands.

Fodder from Forests

Under the forest-based system of livestock production, there has always been enough from the forests to feed animals. In fact, the forests of the country have provided fodder during the season when the grasslands are dry and there is not enough from the agricultural fields to sustain animals. In the Himalayas, herders move their flocks from the upper Himalayas in summer to the lower Himalayan forests in winter. In the Deccan Plateau of south and central India, shepherds herd their flocks towards the forests of the Western Ghats once the rains end.

The many creepers, trees, shrubs and vines of the forests provide the necessary nourishment for animals. Communities living in the forest are the ones who are most knowledgeable about these species. For instance, the Gawli Dhangars of the Western Ghats who traditionally reared buffalo and earned a livelihood selling milk and milk products can clearly identify the best trees to improve the milk yield of their animals. They also know creepers, which are available in late summer when other fodder varieties are scarce. Their knowledge on a whole range of issues to do with plants, their availability and toxicity is quite amazing. While some of this information is available in the codified texts of Ayurveda, much of this is pragmatic knowledge has been gleaned over centuries of use and as yet unknown to formal science.

Fodder from Crops

Under crop-based livestock farming systems, farmers always grew fodder varieties to cater to their animals. References have already been made earlier to the works of Dharampal (1983) as well as Sir Albert Howard (1940) from where it is amply clear that Indian farmers took good care to satisfy the hunger of their animals. Sir Howard, for example, states in his *An Agricultural Testament* (1940) that the "primary function of Eastern agriculture" has been to provide both cultivators and their cattle with food and therefore both 'food' and 'forage' crops were predominantly cultivated. To quote him further:

> "The grass and clover families are widely distributed and cultivated all over the world. From the tropics to the temperate zones and at all elevations and under every possible set of soil and moisture condition. The successful mixed cultivation of these two groups of plants has been in operation for many centuries: In the Orient they were grown together in suitable combinations long before England emerged from the primitive condition in which the Roman invaders found it –an island covered for the most part with dense forests and impassable bogs.
>
> Mixed crops are a rule. In this respect the cultivators of the Orient have followed Natures method as seen in the primeval forest. Mixed cropping is perhaps most universal when the cereal crop is the main constituent. Crops like millets, wheat, barley and maize are mixed with an appropriate subsidiary pulse, sometimes a species that ripens much later than the cereal. The mixing of cereals and pulses helps both crops. A balance between livestock and crops is always maintained. Although crops are generally more important than animals in Eastern agriculture we seldom or never find crops without animals".

Traditional Veterinary Medicine

Ever since animals were domesticated there have always been individuals ministering to the health of animals. In the Indian context, there are two very distinct forms of traditional knowledge. On one hand we find the formal codified system of knowledge in the form of ancient written texts taken from the three systems Ayurveda, Unani and Siddha and the second, lesser known, which goes by the name of folklore which is still practised by millions all over the country.

Anjaria (1984) states that the first veterinary hospital was said to have been set up in the kingdom of Emperor Asoka. Leslie Shirlaw (1940) in an article called 'A short history of Ayurvedic Veterinary Literature', traces the development of Veterinary medicine in India from the Pre-Vedic times up to the 14th century AD. The author states that veterinary medicine in ancient India preceded the development of human medicine and in fact ancient records and writings point to the fact that primitive human medicine was learnt through observing the behavior of lower animals in disease. To quote:

> 'The first literary evidence of the existence of a veterinary cult is found in the Rig Veda (2000-1400 BC) and subsequently references are made in both the "Yajur-Veda" and the "Atharva Veda". It is likely that the physician treated both human and animal complaints. The art developed right up to the 6th century BC after which religious forces led to its decline'.

Shirlaw goes on to describe some of the mythological references made in the *Mahabharata*, *Ramayana* and the *Puranas*. In these texts, veterinary knowledge primarily exists under three main divisions which are horses, elephants and cattle and the famous works of the time. The Shalihotra's *Shalihotram* and Palkapya's *Hastayurved* which deal with horses are mentioned. It is observed that there is no mention of a well known veterinarian or ancient text dealing exclusively on the treatment of cattle and concludes that this is possible because the treatment of cattle was done by the herdsman himself.

There is perhaps another reason why cattle medicine was not codified. Horses and elephants were animals of domination and denoted 'power'. It is possible that texts on these animals were supported by the rulers while those of species maintained by the farmers and peasants were passed on as oral history from generation to generation.

There are very few texts available dealing specifically with cattle. The ones that are known are: *Go Chikitsa* (Anonymous), *Juktikalpatam* (A treatise on draught and other animals by Raja Bhoja 11 AD) and *Pasu Vaidya Mattu Vagadam*, a Tamil book which describes ancient Tamil palm leaf manuscripts. This book details around 250 diseases affecting cattle and their treatments. Diseases include infertility, agalactia, snake bite, disorders of the digestive tract and infectious diseases. In contrast, there are many well known texts on the subject of the treatment of horses and elephants as pointed out by Ramdas and Ghotge (1996).

Unfortunately, these texts have probably not been available to many and cannot really be used easily even today. First of all the language, Sanskrit, and the written word itself were only accessible to individuals who belonged to either the ruling class or to the class who wrote and taught them. It was not accessible to common farmers who could not read. Secondly, no further work was done on reviving these texts in the Middle Ages. It is sometimes even more confusing to place the plants in respect of disease, as plants are known by different local names in different regions while in the texts they are known by the Sanskrit name or often described by their property or properties such as cooling, soothing, fire-inducing, rather than by a name. So the same property or name may refer to different plants in different areas leading to considerable confusion.

On the other hand what has survived, evolved and still lives on is a rich tapestry of folk traditions which have passed down for years from generation to generation, sometimes jealously guarded within communities and tribes. That this form of knowledge has constantly evolved is evident from the fact that plants introduced into the country perhaps only 300 or 400 years ago have been documented in diverse locations with uses quite

---

[1] An Organization working on issues of Livestock Development

distinct from uses in their country of origin such as the use of *Lantana camara* for healing wounds.

In a study conducted by ANTHRA[1] (1997) across 400 small and marginal farming households, it was found that over 85% of them used indigenous knowledge to cure their animals. The major reasons cited by farmers for continuing to use the systems were that people believed in the system and had been using them for a long time. Treatment costs less as it makes use of locally available material; it is effective and works quickly and because the same treatment (plant) may cure many conditions, it is also safe to use and practitioners of this system are present in the village itself.

Traditional knowledge was found to effectively heal certain conditions of the digestive tract, reproductive problems, surgical conditions and skin conditions. Where they did not really prove effective was against infectious diseases.

On the other hand, work by different researchers with livestock herding communities have found extremely effective preventive measures for highly infectious conditions amongst their livestock.

Amongst the Kurubas of Karnataka, there is a traditional method for the prevention of sheep pox. Ramdas and Ghotge (1995) provides details of this method which involves drawing the fluid from a pox vesicle of an infected animal, mixing it with coconut milk and some special herbs and introducing it into non-infected animals by making a cut in the ear. It was observed that this system was sometimes more effective than conventional vaccination methods. The problem with this system was that the wound in the ear was often neglected and resulted in maggot wounds though maggot wounds in the ear are a comparatively minor problem compared to a fatal disease like small pox. Sheep pox is a deadly killer in sheep and entire flocks can get wiped out. Strains of the poxvirus also mutate frequently and vary from site to site. The ovination methods practised by these local healers ensure that the existing strain of virus is used to immunize healthy animals.

Similar prophylactic measures have been recorded for camel pox as has been pointed out by Rathore, Rathore and Rolleffsson (1997). Another interesting practise was observed in Rajasthan where farmers immunise their goats against Contagious Caprine Pleura Pneumonia (CCPP) by crushing a piece of lung of an animal that is supposed to have died of CCPP, mixing it with a herbal extract and then inoculating the mixture with a needle into the large vein of the ear as has been pointed out by the Indo Swiss Goat Project (1993).

While these practices may not be carried out in the most sterile and hygienic manner, the rationale of bringing antigens from a sick animal to a healthy animal is sound and merits further investigation.

There are similar preventive methods for preventing the spread of Foot and Mouth disease and contagious ecthyma (a condition chiefly affecting goats), which involves taking the saliva of an affected animal and introducing it into the mouth of healthy animals. This knowledge though is not freely accessible to everybody and there is considerable specialization within this realm of knowledge.

Village Healers

In practically every village of the country, there are traditional healers who go by different names in different parts of the country. In Rajasthan they are known as *Gunis*, in Assam they are known as *Bez*. They are also known as *Vaids*, *Vaidus*, *Vaidudus*. Interacting with them provides considerable insight into existing knowledge systems.

Traditional healers practise the folk medicine of their region. Their knowledge is usually transmitted to them through their male parent. They have certain strong ethics and codes while practising their profession. Most often they do not charge a fee as they feel the power of their medicine will decrease if they put a price to it. Farmers who have benefited from their skills and medicines are expected to reimburse the healer in cash or kind at a later date.

Healers also do not normally reveal the identity of their medicines. This is partly to protect their own occupational interests but partly also to prevent misuse of various preparations and other malpractices.

Within healers, there is a considerable degree of specialization with some healers specializing in surgical conditions like setting fractures, others in reproductive problems while yet others deal in prophylactic treatments like vaccination. Healers may also specialize in a particular species. Some may treat only cattle while others would treat cattle and small animals. A large number of healers combine human and veterinary care. Healers hold a particular place of respect within their communities as they can cure diseases that many communities still believe is caused either by an evil eye or because the Gods are displeased. Most of the healers met in the course of a study done by ANTHRA (1997) were found to be extremely knowledgeable about local plants and their properties. The uses cited by these healers often corresponded to similar uses in Ayurveda, but it was quite evident that these healers had not studied conventional Ayurveda texts. Healers normally transmit their knowledge to their male children or to one male child. In the absence of a male child, they may teach a nephew but only very rarely do they teach their daughters.

This profession however has lost its healing touch and its magic and not many healers are left today. Most of the healers in the ANTHRA study were over 60 years of age. In this era of miracle cures and 'English' medicine the new generations are not inclined to practise a profession, which does not pay in hard cash. Additionally, threats to the environment and declining biodiversity have meant that the medicines these healers use are themselves getting very rare.

Herders as Healers

The other well-known practitioners of traditional medicine are pastoral communities who have herded animals for generations. This form of specialized knowledge encompasses all aspects related to animal management and health care. The Raikas, Maldharis, Dhangars and Kurumas are all fairly conversant with treatments required for simple diseases affecting their animals. Some of them are also conversant with the anatomy of animals and the changes due to different diseases. They are also aware of the best pastures and the best grasses and fodder for their animals. Many are experts at assisting during the birth of a young one and at caring for orphaned lambs, kids and calves.

Communities living in the forests are naturally better versed with the medicines available in the forest. These may be plants whose bark can be used to set fractures, plant cures for snakebites and other poisons. The medicinal properties of different animals and animal products is also known to them for example the use of the fat of animals such as monitor lizards, pythons and wild boar, the use of the quill of the porcupine and the tail of the skink (a type of lizard).

Traditional Markets

The sale of livestock and livestock produce has always been important activities in the rural economy. Before currency came into existence in the form of coins and notes, animals were bartered for other goods and commodities. The *mithun* still plays the role of currency in northeast India where land and certain items are valued in *mithuns* and land can be bought or sold and paid for with these animals.

Markets were distinct for live animals, for livestock produce and for animal by-products. Many of these occupations have been greatly influenced by the caste structure in the country with specific castes performing specialized roles. A number of these marketing structures exist even today.

In the case of live animals, professional breeders of animals and specific tribes like the Rebari raise animals specifically to sell them to farmers. The sale of live animals is often done in cattle fairs and markets. The now internationally famous fairs of Pushkar in Rajasthan, Sonepur in Bihar, Jehazgarh in Haryana, Malegaon in Nanded district of Maharashtra see some of the largest trade in animals. Different species of animals, cattle, camels, horses, donkeys assemble here and brisk trading takes place. Other accessories needed for animals such as ornaments, pegs, fodder are also available at these fairs.

These fairs also have a religious significance and are held on the grounds of specific prominent temples. They are also held at the same time every year as prescribed by the lunar calendar. For instance, the fair at Pushkar is held around the full moon day corresponding to the lunar month of Kartik and the Malegaon Mela is always on the '*Yel amavasya*' or the no moon day in the winter month of *'Margashirsh'*. Besides these, there are weekly markets held in different prominent villages. Sometimes, there are special days for special species. Raising animals for the market is a profitable occupation and many landless farmers in different parts of the country raise young animals specifically for the market.

The sale of different livestock produce traditionally took place in different ways. Before organised milk marketing came to India, surplus milk found its way to teashops, sweet shops and to households who did not keep dairy animals and could afford to buy milk. Milk was also sold directly by the communities who practised dairying such as the Yadavs, Ahirs and Patels.

Eggs were normally sold at weekly bazaars along with poultry and pigs. Women who raised these animals and birds would take them to these weekly bazaars and sell them. In the case of dung, it was mainly pastoral communities who sold dung and often it was bartered for grain. Surplus wool found its way into wool markets.

Farmers seldom sold meat, and instead special communities of butchers purchased live animals from farmers and these were then slaughtered by the butcher and sold fresh. The by-products after the slaughter of animals such as leather, hides, bones and horns made their way to different users.

In all these activities, religious sentiments and caste have played a fairly important role in defining activities which is why even today many of the traditional marketing structures still exist as they are bound by the rules of faith.

Unfortunately, almost all trade in animals was controlled by middle men and private dealers who systematically cheated producers and consumers alike, resulting in poor profits on the producer side and poor produce on the consumer side as the dealers would resort to all forms of adulteration and cheating to meet their ends.

Women in Livestock Rearing

A section of the community's participation that seldom gets noticed in the long history of the domestication of animals is that of women. While their genuine contribution is seldom acknowledged or more frequently just ignored, nowadays women are targeted by every livestock programme, good or indifferent without the planners really understanding the place women hold in this sphere.

Women have always been working besides men in all livestock rearing activities. Walking with them and the herds in search of fresh pastures, tending to the young and the weak ones when sick, milking dams, spinning wool, weaving clothes and blankets, women have even collected fodder and medicine from the forests, processed the harvest and fed the residue to their animals, collected and sold the eggs of chicken. They have planned wars and intrigues, ridden on horses and on elephants to defend their nations. Above all, they have perpetuated their race by reproducing. So women are pastoralists, foresters, agriculturists and rulers, but just as certain communities have been denied access to resources and have not been allowed to participate equally, women, for different reasons, have been denied opportunities and resources and have not been allowed to participate equally.

Women in pastoral communities

While in a large number of pastoral communities, women travel with the men and their herds, as with the Gujjars and the Dhangars, in certain communities which are semi-nomadic and also practise agriculture like the Kurubas of Karnataka, women mostly stay behind at home with the older members and with the children.

The activity of taking the animals out for grazing is however done chiefly by men and male children above a certain age, often around 8 or 9 years. Amongst the Rebari and the Maldhari herders of Rajasthan and Gujarat, it is quite common to see young boys dressed in the same distinctive clothes as their fathers except many sizes smaller, wielding little canes and helping herd their flocks and herds. Women stay back to look after the young animals, animals which have delivered recently and also to process and prepare livestock produce like milk or wool for home consumption or for the market.

An interesting observation made by Ramdas and Ghotge(1995) was that amongst the Kurubas, the lamb mortality was disproportionately high. On further questioning, it was observed that the ewes were sent out for grazing the day after they had lambed, leaving the young ones in the care of women. The ewes only returned late in the evening, at which time the starving lambs would feed ravenously. This long period of starvation followed by sudden excessive consumption inevitably led to digestive complications and death. When the issue was discussed with the women, they said that having brought up their own children they knew that young lambs needed to be fed frequently but the men would not listen and since they took the decision related to grazing, the women could not really interfere in the matter.

It took two veterinarians (women), and three development workers to convince the shepherds about their mistake and make them change their practices. Pastoral women are largely involved with the processing of livestock produce but men do the selling of the produce. The money invariably remains with men but many groups also maintained that the money was deposited with women for safekeeping. Many of the women from these groups maintained that they were consulted before an animal was bought or sold, though the actual transaction was always done by men. Traditional foods such as tonics and supplements were prepared by women but administered by men. The treatment of sick animals except in the case of young animals was taken care of by men. Recently, many pastoralists have learnt to administer injections as well. This is almost always done by men.

Role of women in the care of animals in tribal society

Amongst different tribal communities, the role played by women again varies greatly. In many of the tribes of the northeast like the Khaṣi, women are almost entirely in charge of their livestock such as pigs and chicken right from their purchase, care and feeding to the selling of stock. As part of its activities, ANTHRA (2000) studied the division in gender roles amongst the Oraon community of central India, which is presented in Table 3.

Here the division of labour was more equitable with both parties

Table 3: The Role of Women in Livestock Rearing Amongst the Oraons of Central India

Women	Men	Children
Releasing animals in the morning and tethering them in the evening	Working animals in the field	
Cleaning of sheds	Building of houses	
Feeding and watering		Grazing
Bringing fodder		
Milking and boiling milk	Milking	
Managing calves	Getting animals crossed	
Administering household remedies, calling the doctor when animals are sick	Administering medicine, calling the doctor	
Giving advice to the men on the sale and purchase of animals	Sale of animals	
Safekeeping of money after the sale of livestock		
Searching for animals which stray or get lost		
	Castration of male animals	
	Slaughter of animals	

Source: ANTHRA : report of training programme held in village Ginabahar, district Sarguja, M.P, May 2000. Submitted to CRS, Bombay.

acknowledging that the others shared in the work. What was interesting to note was that women were actively involved in the decision-making related to the purchase and sale of livestock. Further, women were always entrusted with the money from the sale of livestock.

In a survey amongst the Koya Dora and Konda Dora tribes of the East Godavari district of Andhra Pradesh, it was found that a disproportionately large burden of the daily workload in the tending of animals fell on women. The benefits in terms of cash earned or decision-making and control were seldom within the reach of women. In many communities living close to forests the responsibility of collecting fodder from forests falls on women. As forests are cropped and recede from the villages, women have to walk further everyday to collect enough to feed their animals.

In almost every community, it has been observed that the rearing of backyard poultry is an activity almost entirely managed by women. In many instances, the sale and purchase and the income also remain within the control of women. This money then becomes an important source of income for the woman which she can dispose as she pleases. This money was often seen to contribute to the well being of children.

There are very few women who are specialized as healers or medicine-women for animals. Although they may be aware of a number of plants and medicines, they are never formally trained by practitioners in the village. In the study done by ANTHRA with healers, the number of women trained as healers is less than 4%. Male healers do not train women as a rule. They only teach their daughters if they have no male child and they do not want to share their knowledge with a nephew or any other male relative.

This discrimination has in a major way limited the extremely positive role women could have played. The daily tasks that women are otherwise closely associated with are watering, feeding, tending, cleaning and nursing. A logical extension would have been the knowledge of preparing and administering medicine. Few women know how to fix a fractured bone or perform minor surgeries. They are also not normally involved in delivering young ones despite the fact that in all Indian villages there are traditional *dais* who deliver human babies.

Women in agricultural communities

Amongst the agricultural communities again, there are vast differences in the traditional knowledge of women. Women from castes who have traditionally kept dairy animals are well informed about the care and management of pregnant and lactating cows as well as of young calves. They also have a host of recipes to boost milk yield, cure minor wounds and help digestion, many of the ingredients of which appear from the spice and condiment box in the house.

Women in agricultural households are also largely responsible for post-harvest activities, such as the separating of chaff from grain and the dehusking of pulses. Since women are also responsible in most of these families for offering feed to animals they often know the best and the most relished feeds as well as details of feed intake and amounts refused. Also women are usually the ones who weed agricultural lands, and therefore they tend to be more knowledgeable about the palatability and the relative merits of different weeds that are often used to feed animals.

Women are responsible for cleaning sheds and for clearing dung and urine. Milking is an activity sometimes shared by both, but women most often do the processing of milk in the house to yoghurt, butter and ghee.

Women from smaller and more marginal farms often opt for rearing goats, as they are easy to manage and care for. The entire responsibility and management of these animals usually rests with women. Landless women also prefer to rear a couple or slightly more number of sheep or goats depending on the region. They either take the goat with them when they go to work on the farms of wealthier farmers, or take the animals to a nearby tree where they can graze, or gather the weeds and grasses and bring them home for their animals. Women are seldom ever involved in breeding activities. The reasons for such inequities are largely historical and therefore knowledge of the past often enables us to understand better how certain frameworks and institutions in society have tilted the balance heavily against women. It also gives us directions on how and where we must move on. Not all traditional systems would be appropriate today but they definitely form a base on which future programmes can be built.

4

Emerging Trends in Livestock Rearing

Colonization and its Effects on Livestock Rearing

Rulers and dynasties changed and systems developed and evolved as did rural livelihoods but the major changes in livestock rearing practices came with the advent of colonization when the rulers came from a different continent, a different culture, a different race. We now trace the possible impact of colonization on the different livestock rearing practices described earlier.

The effect of colonization on livestock species of domination...elephants, camels and horses

Even as power changed hands, elephants, camels and horses continued to be the species, which symbolized power, wealth and domination. Emperors and kings of the vast and rich subcontinent rode imperiously astride elephants while their retinue rode caparisoned horses and camels. The Mughal emperors regularly bred and reared elephants. According to Vijaylakshmi and Vaidya (1995), the Maharaja of Nepal, Jung Bahadur, till a hundred years ago owned a thousand elephants. This splendour and pomp appealed to the British when they first came to India. Moorhouse (1985) for example, points out to the number of colourful reports of the Grand Durbar where Maharajas and lesser chieftains came borne on these animals whose accoutrements alone were each worth a prince's ransom in precious metals and stones. The daily allowance in Bengal for a full size elephant as decreed by the British Government was 400 lbs of green fodder viz., grasses, branches of trees, sugarcane or in lieu of the above, 200 lbs of dry fodder. In the state of Madras it was 250 lbs of green fodder or 125 lbs of dry fodder as pointed out by Watts (1972).

The elephant lost some of its glory with the coming of motorized vehicles. It was easier and cheaper in many ways to manage a car than to manage an elephant. Further, with the extensive loss of forest cover throughout the Empire the elephant population were considerably reduced. Elephants gradually disappeared from prominent public profile and retreated into the forests where they continue to be used for lumber activities. Elephants are often not viewed upon kindly by farmers living near the fringes of forests as they stray into fields and destroy crops. This is largely because their own fodder base in the forests has all but disappeared. Some elephants continue to be maintained by temples, especially in South India.

Camels, the other prominent species in the palaces of North India also lost some of their importance with the coming of motorized vehicles but retain their position as animals serving the army in desert stations. The camel is also used as a riding animal for tourist purposes.

Box 4

The Camel in India

India hosts the world's third largest camel population after Somalia and Sudan Rollefson and Rathore (1999). With the exception of a negligible number of two humped camels found in the state of Ladakh, all Indian camels belong to the one humped variety *Camelus dromedarius*.

Within the country the desert State of Rajasthan accounts for the largest number of camels with 71% followed by Haryana with 16%, Gujarat with 4.5%, Punjab 4%, Uttar Pradesh 3.5% and Madhya Pradesh 2%.

In India, the camel is used mostly as a draught animal. In the Thar desert, camel carts are a frequent and a necessary means of transport. There is a reasonable income to be earned with a camel and a cart and development agencies like the Heifer Project International have supported such projects. The camel is used by the Border Security Force of the Indian Government to patrol the border of the country in the states of Gujarat and Rajasthan. In Jaisalmer, the camel is used as a riding animal to attract tourists especially foreign tourists.

There are six well-known breeds of camel in the country:

1. ***Bikaneri:*** As the name implies, this breed is from the Bikaner district of Rajasthan. This breed is mainly used as a pack animal.

(contd.)

(Box 4 contd.)

2. ***Jaisalmeri:*** This breed hails from the district of Jaisalmer. This breed is famous as a riding camel and is therefore known for its speed. They are smaller and of a lighter build as compared to the Bikaneri.
3. ***Kutchi :*** This breed is found in the region of the Kutch in Gujarat. Animals of this breed have a heavy body.
4. ***Marwari:*** This breed is found in the Jodhpur, Jalore, Barmer districts of Rajasthan and is a heavy animal. They are good for agricultural work.
5. ***Mewar:*** This breed is from Chittorgarh, Kota, and Udaipur. This is a comparatively small breed.

   The small size makes it capable of comfortably negotiating the Aravalli hills as a pack animal.
6. ***Double Humped camel:*** These animals are found in the Nubra valley. This animal has two humps. There are very few of these camels left in the country. These animals are mainly used as pack animals in the Himalayas.

The principle camel rearers in the country are the Rebaris, also known as the Raika and are mainly from Rajasthan. Like with the Kurubas of the Deccan, the Raikas believe that Lord Shiva created the first Raika with the express purpose of looking after the camel. The Raikas rear camels to sell them as work animals to other communities such as farmers and traders. The Raikas are again subdivided into two groups: the Godwar and the Maru Raikas. While the Godwar camel rearers are sedentary and engage in village based rearing, the Maru Raikas are usually migratory.

In the extremely arid area of the country bordering Pakistan, the largest herd sizes of camels exist. The camels here range freely through much of the year. They are supervised only during the three months of the rainy season to avoid crop damage. In addition, they need some supervision during the breeding season between November and March. Through the rest of the year, the camels range on their own. Owners manage to stay informed about the whereabouts of their animals by tracing their footprints and identifying them by specific branding marks. A management system of this kind requires minimum inputs as the camels survive in a semi-wild state.

(contd.)

(Box 4 contd.)

> Camels found in the other states of the country such as Punjab, Haryana, U.P, Madhya Pradesh and even Maharashtra are mainly used for assisting agricultural operations. In Maharashtra, for instance, a camel is cheaper than a pair of good draught bullocks and can do the same amount of work. It also does not compete with other animals for feed as it is a browser and relies on tree fodder from trees like the Neem and Babul of which there are plenty in the area. Camel owners here use their animals to transport produce from fields to the main road.

The horse faced a slightly different fate in the hands of the British, a nation of horse lovers. Not only were they important for transport, but sports like polo and riding were very popular and, in fact, horse breeding and rearing was an extremely important activity. Indian breeds however faced competition from more powerful breeds like the *Arab*. But again with the coming of motorized vehicles, horses as a means of transport lost their place. Royal sports like polo and equestrian events continued and have now become fashionable pastimes. Breeding racehorses for profit and sports continues to be a pastime of the rich and the famous.

Indian breeds of horses, which were developed as pack animals and Indian mules, continue to be popular with the army in mountainous border areas as well as with communities who have to transport goods from place to place. However, all Indian breeds of horses as Sahai and Vijh (2000) have pointed out have now been declared as threatened by the FAO.

There are six major breeds of horses found in India.

Kathiawari: From the region known as Kathiawari in Gujarat, these animals are tall slender riding horses whose most distinguishing features are their ears that turn inwards.

Marwari: From the region of Marwar in Rajasthan, these horses were mainly raised for riding and sports. These animals are longer and taller than the Kathiawari.

***Spiti*:** These horses are from the Spiti valley of Kashmir and the adjoining Kulu and Kinnaur divisions of Himachal Pradesh. These animals have been developed as pack animals as well as for riding and can withstand extreme cold, low temperatures and high altitudes.

***Zanskari* :** These animals are found in the Zanskar range of the Himalayas and are used for similar purposes as the *Spiti*.

Bhutia: These are found in Sikkim and are similar to Tibetan ponies and are predominantly pack animals.

Manipuri: These are found in Manipur and Assam and are used as pack animals.

The effect of colonization on crop based livestock rearing

Consequent to the Industrial Revolution, Europe had to feed her machines and her people and the different colonies had to provide the raw materials. The shift in agriculture in India from traditional patterns came with the planting of cotton, indigo, jute and tobacco. These slowly replaced the food and fodder crops of Indian farmers. Wheat also gained prominence, as it was needed to make bread, the staple of Europeans settled in the sub continent. The Occident could not understand the partiality of Oriental farmers for rice, which, in their opinion, was an extremely labour intensive crop. On the other hand, wheat is an extremely demanding crop and exhausts the soil within two years.

Neither did the colonizers have use for the various millets and pulses grown traditionally by Indian farmers as they did not understand them or their uses. Indian farmers were forced to change their intensive practices and adopt more modern methods by the British who felt that farming systems in the Orient were primitive. European farmers used to a temperate climate, temperate region farming conditions and temperate crops tried very hard to bring their farming methods and practices to India, many of which just did not work. What did happen though was that the fertile plains of the country underwent a major change. In *The Organic Farming Source Book*, edited by Alvares(1999), Bajaj and Srinivas, for example, point out to the adverse effects of such colonial practices upon productivity:

> …'in the Chengulpet region, which was one of the earliest in India to come under the British rule, the relatively modest average yields of 2.5 tons per hectare observed in the 1760s had come down to a mere 650 kgs per hectare already in 1788. The yield of lands seems to have persisted around this low level almost throughout India during the whole of the British period. Average production of paddy in India in 1947 at the end of the British rule was less than a ton per hectare, that of wheat around 700 kg, and of coarse grains much below that figure'.

European colonizers could not easily understand the pattern of Indian livestock rearing where animals were raised for manure and draft rather than for meat or milk. They were not familiar with species like the buffalo, which are essentially breeds of the tropics. At the same time, the formal science of 'animal breeding' was gaining prominence and animals were being selectively bred for single production purposes such as meat, milk, wool. This was also possible because the vast untapped natural resources of other colonies like the Americas and Australia could be put to such use. In fact, in the nineteenth century and the early twentieth century, Pagot (1992), suggests that Indian milch breeds like the *Gir* and *Sahiwal* and general utility breeds like the *Ongole*, which was also sometimes called *Nellore* and the *Kankrej* or *Senchor* were taken to the Americas and Australia to be developed as beef breeds.

The imposition of British farming systems and ideas on traditional Indian farming marked the beginning of the divergence between the two systems; livestock rearing and crop rearing which had converged thousands of years ago in the subcontinent.

The effect of colonization on forest based livestock rearing

Before the coming of the British, there were customary restrictions on the use of forests. The hunting grounds of kings were naturally out of bounds for most individuals, as they had to be preserved in their pristine splendour to cater to the sporting and hunting pleasures of the royalty. There were also other forests and preserves that were considered sacred and trees in this region could not be cut. In Maharashtra, these sacred groves are called *Devrais*. In the deserts of Rajasthan, they are called *Orans* and are attached to the village temple. These lands are sacred and animals may graze on the material that has fallen on the ground but trees in this area cannot be cut. These religious measures ensured that community lands were not abused or exploited beyond a limit.

The British soon recognized the commercial value of Indian forests and a Memorandum providing guidelines restricting the forests dwellers' rights over forests was issued in August, 1855. In 1865, an Act was passed which empowered the British Government to declare any land covered with trees or brushwood as Government forests and to make rules regarding their management. In 1878, another Act repealed the earlier one. It divided forests into reserve forests, protected forests and village forests. Severe restrictions were put on peoples' rights over forestland in the reserve and protected areas. In their discussion of the draft forest bill, Hiremath, Kulkarni, and Kanwalli (1995) point out how the British forest policy established in 1884, emphasized the need for state control over forests and the need to exploit the forests for augmenting state revenue. Unfortunately, this affected the people who inhabit the forests, their livestock and their livelihoods in many ways:

✎ The protected forests became the exclusive preserves of the British. Communities who were dependant on them consequently lost their rights to the produce such as fuel, fodder, food and medicine.

✎ The reserved forests were cleared under the British guise of scientific management and systematically planted with species which had neither fodder nor medicinal value such as teak, rubber and tea. As forests came under plantation crops, the grazing lands of animals decreased enormously. Communities from other parts were relocated in the forests to assist the British in their planting operations and they too began to draw on the forests for their sustenance which eventually led to enormous conflicts.

✎ As more and more forestlands came under protected or reserved status, the pressure on village forestlands grew enormously, the backlash of which is evident even today.

✎ Many forest communities practised slash and burn agriculture. As the land available for this kind of agriculture became less, the cycles became shorter. Where people used to return to the same lands after 20 years, they were forced to return to these lands now in 5–6 years. This of course had a deleterious effect on the regeneration of the forest. It also had a deleterious effect on the crops, as the fertility of the soil would not have been restored within the period. Livestock in this region suffered along with the people as wild fodder species available from the forest as well as crop residue available from millets grown under the slash and burn system decreased. Natural selection ensured that the hardiest and toughest breeds of animals and species that could withstand the pressures of a low plane of nutrition only managed to survive.

The effect of colonization on pastoral systems

It is perhaps more difficult to establish a direct link between the effects of colonization and pastoral systems. First, pastoral systems were practised in the arid and semi-arid parts of the subcontinent where it was hard to practise agriculture and where forests were scanty or of the scrub type. Incidentally, most of the semi-arid and arid areas were never brought under direct British rule, but continued under local rulers or protectorates. The princely states and agency tracts of Kashmir, Himachal, Rajputana, Saurashtra, Hyderabad and Mysore fall under this category and these areas map directly onto areas where the arid and semi-arid parts of the subcontinent lie. These are also the regions where the maximum number of pastoral communities continue to reside. Interestingly, these are also the areas where some of the better breeds of livestock also exist. There were migrations from agency tracts into regions where the British ruled, for example, from Rajputana into the neighbouring Punjab and Sind, but pastoral systems have continued. This is probably because these systems are the most fluid and dynamic of herding systems and respond very quickly and variably to changes in the environment. While in good years, herders will increase their flock size, on the other hand during bad years, they reduce their flock sizes by selling or migrating with their animals to areas which have fodder. Sometimes, they even change the species they herd.

Institutions Created by the British

Not everything the British colonizers did was damaging or detrimental. To their credit, they set up veterinary colleges, various research stations, and conducted research on various Indian practices. Most importantly, they documented everything they saw or did in various written texts, which are available today and provide valuable insights into practices of that period in history.

The British were also responsible for creating the Royal Commission on Agriculture (RCA), the Imperial Council of Agricultural Research (ICAR), the Imperial Dairy Research Institute (IDRI) and the Imperial Veterinary Research Institute (IVRI) that were all to have an effect on future policies set by Independent India.

By 1927, the Royal Commission on Agriculture in fact concluded that the primary function of cattle in India was to serve as draught animals. Milk production was thought to be a secondary consideration. The Commission gave feeding priority over breeding. As significant additions to grazing areas were not possible, the need to raise the productivity of grasslands was stressed.

However, the world wide economic depression of the thirties followed by World War II probably resulted in a complete change of priorities, so that India at independence also inherited a legacy of unresolved problems.

Independent India: Policies and Practices Affecting Livestock Rearing

While Independence was a dream, which finally came true, there were scores of problems, which the new nations leaders had to face. Two corners of the country were ravaged and torn by partition. There was a high level of rural indebtedness compounded by food insufficiency and a low per capita of food availability. Bengal had just passed through a very bad famine. There was an urgent need for land reforms. While one might wonder how these diverse events affected the livestock in the country, they did in surprising ways, which the decision-makers in their wisdom probably never imagined.

The effects of emerging policies on crop based livestock rearing systems

The 'Grow more Food' campaign was the first step the new nation took to achieve food security. The production of food grain had to be increased very quickly. Regrettably, this soon turned into a grow more wheat and grow more rice campaign and in the process different millets and pulses were totally ignored. In comparison to a crop like jowar (sorghum) which was grown rather extensively throughout India both as a food and fodder crop, wheat and paddy crop residues fare rather unfavourably as fodder. The system of multicropping such as growing jowar together with multiple pulses like horsegram *(Dolichos biflorus)* and cow pea *(Vigna ungiculata)*, all of which are important food and fodder species have all but disappeared and an era of monocropping with wheat and rice as the chief focus was soon ushered in.

Over the years, even the varieties within wheat and rice have disappeared and monocultures of bakery-useful wheat and *basmati* rice are now increasingly being grown throughout the country. *Sarvati* rice, the red grained rice grown in the pocket-sized paddy fields of Ratnagiri in Maharashtra and *kulith* or horse gram *(Dolichos biflorus)* were important crops in the region. *Sarvati* rice is supposed to be medicinal for both animals and humans and any animal farmer in India would swear that horsegram is one of the most nourishing foods especially for bullocks. Unfortunately, farmers in this area no more want to grow these crops as they find that the season for harvesting these crops interferes with the harvest of new varieties of rice which are being promoted by the government in an effort to increase food availability. Consequently, the livestock in the area have lost out on an important source of protein fodder as well as a medicine. In effect, the supply of food to humans seems to have been achieved at the cost of a serious loss in the quality of nutrition for farm animals.

Independent India also welcomed what is popularly known as the Green Revolution. This revolution brought with it an era of improved seeds and hybrid varieties. While efforts were made to promote species, which were useful as both food and fodder, what ultimately did take off were species of dubious value. One of the species has been the hybrid jowar. While it apparently can be grown on marginal lands and has greater grain yield, the stalks are soft, hollow, inedible and fares rather poorly when compared to traditional varieties of jowar, which are reputed for their sweet succulent stalks.

The Green revolution initially sought to create surpluses in the well-endowed areas of the country: The fertile areas of the different river basins and the irrigated areas. This therefore necessitated the introduction of mechanization to be able to counter the increased demands of production. As tractors, disc ploughs, threshers and harvesters entered the scene, they gradually replaced our draft animals especially male cattle. However, there was also an excess of local crop residue available in the area through increased production which had to be utilized. The obvious choice was the buffalo, which could consume strawof fairly indifferent quality and still produce milk of high fat content, more in fact than the cow. Further, the buffalo could be stall-fed; it does not need to be exercised and does not require great care and management as long as there is a pond nearby to wallow in. The buffalo thus came to be the species of choice in Punjab, Haryana and western Uttar Pradesh. Soon there was more crop residue, less cattle, more buffalo and more milk that in fact seemed a highly desirable combination. This in fact ushered in another revolution in the country called the White Revolution that was also propelled by a host of other critical factors which occurred at that point of time.

Buffalo rearing was therefore then promoted as a key solution to all poverty development programmes. Buffalo based dairy programmes were designed wherein North Indian buffaloes, *Murrah*, *Mehsana* and *Surti* could be distributed to landless women, who could buy fodder from the increased crop residue in the area. They could then sell milk through cooperatives and finally be alleviated from poverty.

Box 5

Important Dairy Buffalo Breeds of India

Murrah

One of the most popular breeds in the country, it earns its name from its curled horns. The home tract is the state of Haryana though *Murrah* buffaloes have been distributed throughout the country. The animal is massive, jet black in colour and is an excellent milk producer.

Nili Ravi

From the valley of the Sultlej and Ravi rivers, this animal is distributed throughout West Punjab. The Animal often has white markings and many of these animals lack pigment in the iris of their eyes. This breed is also an excellent dairy animal.

Surti

These animals are from Surat, Kaira and Baroda districts. Smaller in size than the *Murrah* and the *Nili Ravi*, these animals are distinguished by their sickle-shaped horns. Although their average milk yield is less than that of the *Murrah*, they are very popular in areas where fodder is not as abundantly available.

Mehsana

The home tract of this breed is Mehsana in Gujarat. Some reports claim they are a cross between the *Surti* and the *Murrah*. Mehsana animals are known for their regular breeding. These animals are also good dairy animals.

Jaffarabadi

This breed gets its name from the town of Jaffarabad in the Gir forest of Gujarat. These animals have large horns that curl downwards encircling the face. Raised by Maldhari herders, these animals are also excellent milkers.

Bhadawari

This animal is found in the Agra and Gwalior regions of Uttar Pradesh. The skin is copper-coloured. Although the average milk production is not very high, the milk of this breed has a very high fat percentage 10 – 13%. Most other breeds average 7 – 8%. In areas, where this breed is popular the production and sale of *ghee* have become important activities.

Besides these there are breeds like the *Pandharpuri* of Maharashtra, the *Toda* of the Nilgiris, the *Sambalpuri* of Orissa which are not essentially dairy animals.

These programmes however failed to perceive certain ground realities. What works in the well-endowed river valleys need not necessarily work in areas, which are not so well-endowed. Mechanization and buffalo rearing are being promoted rampantly in the semi-arid areas of Rajasthan, Maharashtra and Andhra Pradesh. The buffalo which does so well in the farms of U.P, Haryana and Punjab, especially the big breeds like the *Murrah* and the *Nili Ravi* produces up to only a half or a quarter of its potential in the small farms of Beed in Maharshtra or Medak in Andhra Pradesh. First, on account of the chronic scarcity of water in these drought-prone areas, there is insufficient fodder. Secondly, there are not enough ponds to wallow in. This leads to a host of reproductive problems in the buffalo especially a prolapsed uterus after delivery. This in turn leads to large expenditure on medicine and veterinary care. Thirdly, with the shift in production to paddy, cotton and other crops and the decrease in traditional fodder crops, the crop residue is not nutritive enough and animals fail to conceive on time. Finally, because these animals have been distributed to communities who have never managed these kinds of animals before they are at a complete loss when disease strikes. Unfortunately, there is hardly any training given to these women prior to initiating them into the activity. With high costs of production, low milk yields, poor rates of conception, it is little wonder that these programmes fail after some time.

The introduction of machines has caused yet another problem. Mechanization is now being used in post-harvest operations, a task that traditionally women used to do. Threshers and combine harvesters are not as discriminating as women on the farm, who carefully separate fodder from grain or pulse. The machines cut up the stalk into particles which animals find difficult to consume.

Most importantly, unlike draught animals, tractors do not fertilize the soil. This has meant the farmer has to add huge quantities of fertilizer to keep production levels up leading to an increase in the cost of production. Thus from what was a fairly closed production system under traditional agricultural systems where animals and agriculture complemented each other, modern agriculture has been forced to face up to the realties of an international market economy.

As the number of animals in the prime agricultural areas have decreased, and the demands on production have increased, chemical fertilizers have had to be introduced. To be able to purchase chemical fertilizer for their fields, farmers have had to grow cash crops like spices, grapes, flowers and chillies which do not necessarily have fodder value. Farmers have to be careful as these fields are not encroached upon or grazed upon. Animals are not liked in orchards and slowly the land that once welcomed animals has slowly begun to reject them.

Unlike natural manure, chemicals and synthetic compounds cannot replace the organic carbon content of soils. Many cash crops exhaust the soil quickly. At some future date, farmers who do not have sufficient animals or who have sold their animals will be forced to buy organic manure from some other source to revive and rejuvenate their soils.

Traditionally, oil seed cakes were used to supplement animal rations. However, policy measures recommended the sale of oil seeds themselves. Once we began to export our oil seeds there was insufficient oil seed cake available and another critical ingredient in the diet of animals got in the process reduced.

On the other hand, urea, offered as subsidies to farmers was available in plenty and so was molasses from sugar factories. The next discovery was the urea molasses brick to supplement rations. The natural diet of animals slowly began to get replaced with chemicals and synthetic compounds.

Today in many parts of western Maharashtra, the most agriculturally prosperous part of the state, middle and small farmers who have shifted to modern farming systems grow cash crops and vegetables to suit the urban market. Since they do not have adequate fodder, they have reduced the number of animals they keep. They do not realize that fossil fuels are getting more and more expensive and there are no 'modern' energy solutions to replace them. They have sold the one solution they had, draft animals. They also do not know how to predict the market and blindly follow their more wealthy neighbours. They live under a myth that they have progressed, but the bitter truth is that if tomatoes do well one year, everybody rushes to plant the same crop the next year and the market price falls.

As the demands on land have increased, farmers have also begun to crop in all seasons as well as in soils and in areas not really suitable for a particular type of crop. Non-seasonal crops run the risk of greater pests infestation. While much has been written about the ill effects of pesticides on humans, there is comparatively little mention of the contamination of crop residues with pesticides. Development officials and agriculture officials blindly promote the use of pesticides to farmers without understanding the downstream consequences of such rash promotion. Farmers and others handling pesticides are careless in the way they handle and store their chemicals. They are not adequately warned about the harmful effects of these compounds. Worse still is the fact that while we may wash our grains before we consume them to get rid of harmful chemicals, we cannot and do not wash crop residues before feeding animals.

Pesticides are also contaminating water sources from which both humans and animals consume water. Buffaloes often wallow in village ponds, which are contaminated with pesticides. These pesticides are absorbed by the skin. The contamination of milk with DDT, a banned pesticide also continues to be a major health issue.

Deaths of animals due to the consumption of pesticides seldom gets reported because even local veterinarians do not really know the exact symptoms and find it difficult to diagnose and distinguish cases when they arrive. Moreover, these cases cannot be treated by village healers either. Even when cases are identified as being related to pesticides they seldom reach the proper authorities where steps to remedy the damage can take place.

Reforming the land

While the Green Revolution and its effects were primarily adopted in the river basins and the irrigated tracts of the country, an attempt was made to enforce these measures in other areas too.

While land reforms were required and redistribution of land was promised, it has never really been achieved in the country except in a very few states. Instead the government has taken the softer option of giving so called marginal lands to communities who have demanded land. Inevitably, these lands have been village grazing lands. This has happened in quite a few states, notably so in Rajasthan and in Marathwada in Maharashtra.

Poor farmers who do not have the resources to use machines on their farms manage the grazing lands that now come under agriculture. They do not have the means to buy quality draft breeds or fertilizer either. As land sizes are small, they resort to stocking nondescript animals to help them with their agriculture. As the net returns from these animals are low, they are forced to stock more numbers.

However, these farmers in the absence of irrigation and water are not in a position to grow fodder to sustain their animals for the entire year. Animals continue to depend on common resources like forests and grazing lands for fodder. As these have reduced in area or got fragmented, the grazing pressure on the available lands has increased.

Irrigation facilities were provided to increase the country's food security and this automatically should have ensured fodder security, but unfortunately during the course of time food crops have been replaced by different cash crops, which feed into an urban market. Improper irrigation methods have also led to problems of the soil like water logging and salinity, which are seen in large tracts of Punjab, Haryana and Rajasthan. This in turn results in reducing the productivity of the land and resulting in lower availability of food and fodder.

The unresolved paradox

Small farmers through various credit systems, state subsidies and programmes are being encouraged to turn to monocropping in preference to mixed farming systems. As the diversity of their lands decreases and productivity goes down they become even more impoverished. The paradox of the problem is that they then turn to livestock farming as it provides quick returns. They often stock more cattle and even goats which further puts pressure on community grazing lands. Sometimes they even take up pastoralism or choose to migrate. But as there is insufficient grassland and fodder, pastoral communities who were dependant on nearby grazing lands for their animals move to new areas leading to new tensions and conflicts.

Which comes first... the goat or the desert?

As the grazing resources decrease, not only does land get overgrazed, the better varieties of grass disappear and in its place the less palatable and coarser varieties of grasses remain. It becomes impossible to rear the better and more demanding breeds or species on such resources, so livestock herders then opt for smaller and hardier breeds and species. Herders also prefer to stock species of animals, which can survive long periods of drought and acute fodder shortages. The ideal species under these circumstances is the goat. The Hanumanth Rao Commission appointed by Shri Rajiv Gandhi discovered that the goat followed the process of desertification and not the other way around. Yet the perception persists that goats damage the environment rather than that they are amongst the few species capable of surviving in an already damaged and degraded environment.

This perhaps to some extent explains the fact that despite the very cautious approach the government has taken towards promoting goats, the population of goats has doubled in the past 50 years. In fact, if one were

Fig 4.1 Livestock Population in Rajasthan in 100,000's

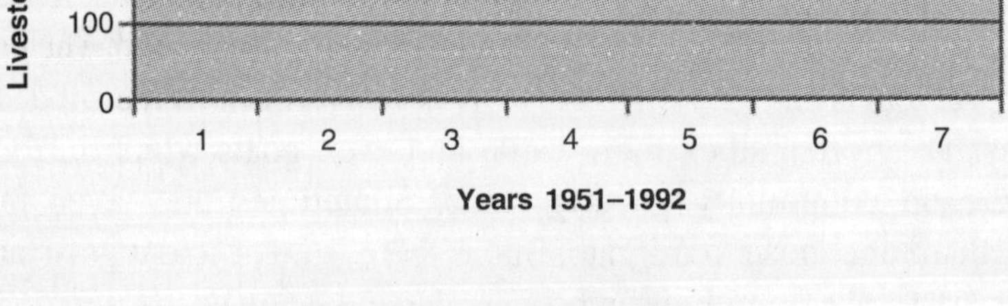

to view the changes in livestock stocking patterns in Rajasthan one would see distinct peaks and troughs. The peaks correspond to years with good fodder where the livestock population increases while the troughs mark years where there have been droughts and farmers have reduced their flock sizes or animals have died due to fodder stress.

Unresolved conflicts

As the pressure on grazing lands increases, herders are forced to turn to the forests for feeding their animals. The forest policy framed by the British continued into the post-Independence era. The protected areas were declared national parks and sanctuaries to protect the country's wildlife. Reserve forests continue to be planted with various commercial and quick growing species like teak (*Tectona grandis*), *Acacia auriculoformis*, and eucalyptus, all of which have little or no fodder value. The village forests lands already considerably degraded with various kinds of pressure have been rendered almost barren.

Illogically conceived and randomly implemented policies have also affected and politicized the issue. Previously, grazing fees used to be levied and herding communities respected and paid them. Somewhere along the way these fees got abolished and herders stopped paying them. When attempts were made to revive grazing fees, this has almost always led to controversy and conflict with the issue largely remaining unresolved.

One of the reasons stated by farmers who live near the Aravallis in Rajasthan for the degradation of forests in their area is the indiscriminate grazing by animals. They maintain that grazing fees that earlier used to be charged are no more charged. Yet, when the states of Madhya Pradesh and Maharashtra charge fees from herders from Rajasthan when they bring their herds into these states, the herders complain loudly.

The fault does not lie with the State alone. Various well-meaning agencies over time have wrought considerable damage. As the forests got degraded, various attempts were made at greening, increasing the biomass availability, preventing the process of desertification and providing fuel. In the process, certain key species got selected. The *'Prosopis juliflora'*, known variedly as *gando Babhal* in Gujarat, or the *'vedhi Babhal'* in Maharashtra or the *'mad Babool'* to the rest of us, is one of them. This species was planted in large quantities all over Gujarat and Rajasthan. It is said that the aerial seeding of the plant was done with the blessings of the erstwhile Maharajas of Marwar and Kutch. Today, villagers are regretting the programmes, which promoted the scheme as it has taken up all village grazing lands and common lands. The *'mad Babool'* is not edible; it grows wildly as its name implies and prevents other species from growing under it or with it. The shade it provides makes an excellent breeding ground for mosquitoes and even goats that eat almost anything get blisters in their mouth after eating it.

In this way different agencies, different departments and the state have begun the process of alienating animals from their land as the process of development and change spreads its tentacles on unsuspecting victims. A quick glance at the effects of policies from departments, which are not directly concerned with livestock, reveals the following change in patterns:

i. A shift from cattle rearing to buffalo herding in irrigated areas.

ii. Disappearance of important draught breeds of cattle.

iii. Shortages of fodder: variety, quality and quantity.

iv. Enormous increases of nondescript breeds especially in marginal areas, which are hardy but produce very little.

v. Increase in conflicts between now competing systems, pastoral, forest based and crop based.

Let us now examine the policies of the Department that directly deals with livestock and their effect on the livelihoods of animal farmers.

What the policies say

The very first attempt to conceive a set of policies for livestock development for the nation as a whole was the Royal Commission of Agriculture in 1928. The Commission clearly recognized the role of draught animals, the problem of excessive cattle population and the scarcity of feed and fodder. The commission linked the vicious cycle of low quality stock, fodder scarcity and excessive numbers to the Hindu religious sentiment against the slaughter of cows.

After independence, livestock policy came under various national plans, which are described as below:

Thrust Area Projections: Achievements and Failures of the Policy 1950-2000

Year: 1951-56 and 1956-61
Plan: 1st and 2nd Five Year Plans

Trust Areas and Projections:

- To Increase milk production
- To improve milk supply to urban areas
- City owned cattle to be brought under peri-urban co-operatives supplying to Government dairies.
- To improve quality and supply of draft animals.
- Grading up of indigenous breeds to improve milk supply.
- Fodder scarcity to be tackled.
- Key Village Schemes to be launched
- Programme of castration of Scrub bulls to be taken up
- Artificial Insemination programmes to be taken up and breeding of livestock to be regulated.
- Diseases of animals to be controlled

Achievements
Milk production increased and milk supply to urban areas improved. Investments were made in veterinary hospitals and dispensaries.

Failures
The quality and supply of draft animals could not improved nor could the grading up of indigenous breeds to improve milk supply. The problems of large numbers of urban cattle continued. Fodder scarcity remained a problem.

Year: 1961-66
Plan: 3rd Five Year Plan
Trust Areas and Projections
The thrust areas were essentially the same as the previous two plans. Additionally there was a need felt for an alternate policy for breeding cattle for increasing production. Other focal areas were the setting up

of feed mills for balanced feeds, promoting the capabilities of indigenous manufacture of dairy plant and machinery. Another focal area identified was animal by products skins, hides and fallen animal products.

Achievements
Some headway was made in the setting up o feed mills for balanced feeds and also in promoting the capabilities of indigenous manufacture of dairy plant machienery.

Failures
Inadequacies of the Key Village Schemes were realised. Fodder needs could not be addressed. Urban milk schemes failed and there was total failure in implementing policy on draft animals.

Year: 1969-74
Plan: 4th Plan delayed by three years

Thrust Areas and Projections
While most of the above continued, there was added emphasis on milk production in rural milk sheds and movement of processed milk from rural areas to urban demand areas.

Achievements
The setting up of NDDB and the launching of Operational Flood!
The total milk production of the country increased considerably
Gift commodities of milk powder from external agencies were used to end commodity aid and dependence
Cross breeding with exotic milch breeds gained momentum.
Simultaneously buffaloes established there preeminence as the mainstay of the Indian dairy Industry.

Failures
Draft animals and fodder issues continued to be unresolved

Year: 1974-1990
Plan: 5th 6th and 7th Plans

Thrust Areas Projections
- Implementation of operation flood
- Investments in dairy plants

- Milk processing Plants
- Linking rural small producers to urban markets
- Fodder develoment schemes, distribution of mini kits and fodder seeds
- Need to look into alternate strategies for feeding
- Expansion of capabilities of department to deliver curative services to cattle and Buffaloes
- Establishment of National Rinderpest Eradication Programme.

Achievements

Progress was made in the National Rinderpest Eradication Programme.

Failures

Fodder crisis continued acerbated by drought in the arid and semi arid parts of the country. The Government could not really come up with a coherent plan for handling fodder shortages in non irrigated areas. The Government was also unable to stop the entry of buffaloes in to periurban and urban areas. Negative selection of buffaloes continued with quality animals being rejected after a single lactation.

Year: 1992-97

Plan: 8th Five year Plan

Thrust Areas and Projections

Remain essentially the same as above

Achievements

IRinderpest Eradication Programme continued. There was also considerable growth in the number of institutions catering to curative veterinary services.

Failures

The problems of fodder scarcity continued to get worse. The number of quality draft animals went down drastically. The state was really unable to provide quality preventive services to cater to the needs of existing and emerging infectious diseases.

Other aspects of the policy

- Cautious approach to goat development. The government by and far has not encouraged programmes with goats. However, the goat population has increased by 100%.

- Despite major incentives to sheep especially wool breeds, the sector has stagnated.
- Plan allocations inadequate to make any impact.
- Neglect of buffalo breeds and with the emphasis on dairy farming using buffaloes farmers, especially dairy farmers, in the urban and peri-urban areas procure new animals every year when they are pregnant. Use them through the period they are in milk which is up to about thirteen months and then sell them to slaughter houses. They repurchase new animals and use them. Similarly, in many areas, male buffaloes which do not have any use in dairy farming are abandoned and left to die.
- Under the third and fourth five year plan, skins, hides and fallen products improved due to policies such as ban on export of raw hides, restriction on export of semi- finished leather and incentives for finished goods.

According to Dr. MPG Kurup (1999), an acknowledged expert in the livestock sector in India and the Coordinating Consultant for the Steering Group of the National Livestock policy perspective, while the government made enormous strides in the dairy industry there has been comparatively little change in the other sectors. He adds that as the cooperative network under Operation Flood and price incentives moved in to provide the much needed market stimulus, the cross-breeding of Indian breeds of cattle with exotic European cattle, as a national policy gained momentum. Milk gained recognition as a marketable commodity. The producer price offered by the co-operatives, in effect, became the support price for milk, and the pre-eminence of buffaloes as the mainstay of the dairy industry got established.

He goes on to state, that was insufficient policy focus on the issue of draught animals. As a result of neglect of the policy plans spelt out in the Second and Third Plan periods for selective breeding of indigenous draft breeds and grading up of non descript animals with Indian breeds, we have failed to built up a superior stock of work animals.

Fodder development has also been marginalized despite growing concern over its acute scarcity. Although the devastating effect of the two year stretch of severe, almost country-wide drought during 1985-87 focused attention on the need for alternative strategies for feeding livestock, but not much progress was made in fodder development. Unfortunately, the Seventh Plan could not rise above the department approach of fodder development schemes, mini kits etc which anyway are only applicable for irrigated areas and not for drought-prone as well as arid areas where, in fact, livestock rearing is a major livelihood activity.

The thrust of the State Animal Husbandry Departments has mainly been on curative treatment rather than on prevention and control. Consequently, despite expansion of the capabilities and coverage they have failed to achieve their targets. Foot and Mouth disease continues to be a major problem. The increase in allocation made available under the successive plans to enlarge the scope and size of the veterinary institutions, has largely been for free curative veterinary services rather than for disease prevention and control, an area which lies exclusively in the public interest.

Despite the recommendations of the National Commission on Agriculture to restrict and limit the goat population of the country, the growth of the population from 1951–1950 was almost 100%. On the other hand, in spite of a favourable policy environment and the many promotional schemes, the population of sheep grew very little over the four decades of planned growth. Sheep have been promoted exclusively for wool production, and policy measures, research and development efforts also reflect this bias though considerable portion of our sheep are, in fact, meat breeds.

What the policy does not state and what gets implemented in practice

- That cross breeding programmes with exotic animals has become the major focus of the Animal Husbandry department almost to the exclusion of other programmes and a State Animal Husbandry Department is evaluated by the success of its cross breeding programme. For instance, a State like Kerala where a large percentage of cattle are cross breeds as opposed to local breeds is considered to be a success story While cross-bred animals might perform well in well-endowed areas with adequate inputs of fodder and veterinary care, they fare rather poorly in marginal, fragile and less endowed areas where fodder is scarce and veterinary support is weak. Yet, the Government has sought to promote these schemes uncritically.

- For a cross-breeding programme to be effective, it is important to maintain pure lines of good local breeds. This did not happen. Breed records and pedigrees were badly maintained in most villages. In effect, the programme became a breed up gradation programme of nondescript zebu cattle with high milk-producing exotic cattle. After the third filial generation, the cross- breds were neither hardy tough *zebus* who could withstand tropical conditions nor good milk producers but a sorry, even more non-descript variety of hapless cross breds with the combined defects of both breeds (Table 5). One has only to visit some of the poorer hilly areas of the country to see this unfortunate looking animal, which 50 years of independent India has created.
- Artificial insemination with exotics was given such a large incentive that traditional systems of breeding collapsed. Communities who raised male animals for the purpose of breeding could not face the competition of almost free services given by the Government. In the process, many good dual-purpose breeds like the *Deoni* from Maharashtra, *Ongole* from Andhra and the *Gir* from Gujarat have lost out.
- While many in the field of Livestock Development keep talking about preserving local breeds and the state government and central government on paper write extensively on this subject, the truth is that there are very few good animals really left in farms run by the government. A few *goshalas* and *panjra* poles in the States of Gujarat, Rajasthan, Madhya Pradesh and Maharastra maintain herds of animals found in their region, but in other parts of the country there are hardly any institutions or resources forthcoming that can help preserve these breeds. The *Ponganur* of Andhra, the *Vechu*r of Kerala, the *Dangi* of Maharastra, are breeds which have excited a lot of interest lately because of the interesting and unique features they

the 'milk capital' of the country. We were proudly told by the dairy cooperative that there were none. Later, we did some independent research and found that there were over a 1000 goats in the village. More in fact than the number of buffaloes in the village. While the buffaloes were reared by comparatively well-to-do farmers, the invisible communities of tribals, landless and nomads reared the goats. In essence, the wheat and rice revolution in agriculture has been paralleled by the cross-bred cow and the buffalo revolution in animal rearing.

- Despite the fact that there has been a mushrooming of veterinary services in the country and in many places an over-supply of veterinarians, poor people in the country do not have access to proper veterinary care and resort to traditional systems and superstitious beliefs when there is a crisis. Posts for veterinarians in veterinary clinics in remote areas remain vacant for years.
- That although the role of women is acknowledged on paper as being significant to the livestock industry, women are still kept out of crucial decision-making roles. Most programmes, in fact, only increase the burden on women.

Land Grant Institutions and Beyond

When the British left India, they left behind an Imperial Council of Agriculture at Pusa near Delhi which became the Indian Council of Agricultural Research, the Imperial Veterinary Research Institute at Mukteshwar and Veterinary Colleges at different places including the cities of Bombay, Madras, and Calcutta. However, the country that India looked towards at that time for guidance was the U.S.A, which emerged triumphant and unscathed after the Second World War that had crippled Europe. American research establishments were flourishing and paving the way for exciting new research and our national leaders looked up to the U.S.A for technology and aid. Little did they realize that the continent of North America was vastly different from a country which had been under the plough for thousands of years.

Box - 6

The Taste of India

Till the Europeans arrived in India, organized dairy farming did not exist. Communities like the Yadavs, Ahirs and Patels kept dairy animals and sold milk often to tea and sweet shops. But, in fact, in many parts of the country, amongst certain communities, the sentiment was *Doodh bechna aur poot bechna samaan hai*, which means selling milk is equivalent to selling your own son. This sentiment still survives in some places.

The European colonies, however, needed milk and the organized sale of milk began in the period when the British ruled. One of the first large-scale dairy farms was established by the British in Allahabad in 1891. In 1941, the Imperial Dairy Research Institute was established in Bangalore.

Meanwhile, further north in the Kaira district of Gujarat, dairy farming had established itself as an important activity amongst the predominant landowning community of Patels and large quantities of milk were available. Different enterprises began in this region to make use of the surplus milk. In 1900, Kaira saw its first dairy plant built by a Swedish gentleman. In 1911, a German started manufacturing casein in the same region. Subsequently, a dairy farm was established in Nadiad by an Englishman while in 1929 in Anand, Mr. Polson, a Parsee entrepreneur installed a modern dairy plant to manufacture butter, cheese, casein and milk powder. Many of us may still remember the brand name of Polson butter.

In 1946, however, the Patels of Kaira united to form a Cooperative, the Kaira District Cooperative Milk Producers Union Ltd. to sell milk directly to the Bombay Milk Scheme. The Cooperative had the blessings of a number of very prominent national leaders including, Sardar Vallabhai Patel, Shri Morarji Desai and Shri K.M.Munshi. The Anand cooperative structure consisted of a two-tier system, the Village Milk Producers Cooperative Societies at the village level and the District Cooperative Milk producers Union at the district level. The organization of milk producers was owned by farmers and is better known now as the famous 'Anand' Pattern. The Anand pattern was so successful in its home district that other states and districts as well decided to replicate it. Operation Flood

(contd.)

(Box - 6 contd.)

I, the largest dairy development project in the World was launched. The breeds of animal behind this enormous success were the *Surti* and the *Mehsana* breed of buffalo.

In the period between 1970 and 1985, the recorded milk production rose dramatically from 22 million tonnes to 44 million tonnes. By the mid-nineties, India had established herself as the world's largest milk producer. The Anand Milk Producers Union Ltd. (AMUL) model had come to stay.

Operation flood, however, has excited much controversy. The circumstances that helped establish AMUL and other dairies in Gujarat were not there in other parts of the country. The cooperative spirit and movement were absent in many areas that led to the failure of this model in some states. There were also criticisms on the cross-breeding efforts advocated by the programme as well as the excessive emphasis on marketing because of which it was felt that milk-producing households sold all the milk they produced thereby denying their families, especially their children of this valuable source of nutrition. On the other hand, the urban cities and metropolises are now flooded with milk and milk products, many of which are new to the Indian palate. Cheese, table butter, milk powder and ice creams, besides *ghee*, *shrikhand*, *mishti dahi* and *gulab jamun* are now available all year round for those who have the money to buy them.

Table 5: Total Milch Animals and their Daily Output during 1997–1998 in Maharashtra

Breed	Milch animals in lakhs	Per day production peranimal of milk in kgs
Indigenous	62.17	1.496
Cross bred	14.15	6.786
Buffalo	42.09	3.557

Source: Govt. Of Maharashtra, Dept of Animal Husbandry

First, the USA was a country where large-scale agriculture had begun only very recently. Even today less than 5% of America's population is farmers. On the other hand in India at least 60% of the population are in the rural areas and depend essentially on farming for a living.

The U.S. had vast unexploited natural resources. Theirs was a temperate climate with temperate tastes. Consequently, the main foods and cereal crops were different, the livestock reared were different and the reasons for reasing livestock were different. After all, the *Gir* and *Sahiwal* dairy cattle breeds of India left for American and Australian shores as beef breeds. Yet the American model looked attractive. It presented a picture of fat, contented animals on acres of lush pasture cheerfully producing vast amounts of milk, which could be converted to healthy nutritious dairy products. It presented acre after acre of golden fields ripe with wheat and corn. It was indeed an appealing image and the path was paved for land grant universities modeled after land grant universities in the U.S. to enter India.

Hisar in Haryana in the heart of the Green Revolution belt is a typical example of such a university. The beautiful campus modeled after an American university has six colleges, which impart education in the veterinary and animal sciences, home science, basic sciences, agriculture and sports. Students are encouraged to take up interdisciplinary courses. The facilities for research are fabulous and a large portion of the research in the veterinary college is on dairy buffaloes and cattle. Hisar probably gets a considerably larger chunk of funds for research than the other veterinary or agricultural universities in the country. Visiting dignitaries from other countries are driven to this campus, which is located less than 200 Kms from the nation's capital, New Delhi.

The education imparted in these universities is based on text books written in Australia or the U.S. After studying these texts, students are left with the impression that traditional systems are inefficient, local breeds are poor producers and almost everything traditional in agriculture and livestock farming systems is backward. A point reinforced by the fact that Indian farms present a sorry picture contrasted with the American dream.

Animal science as a subject is taught both in veterinary colleges and in agriculture colleges and graduates from both disciplines can specialize in subjects that concern livestock production, including breeding, nutrition, production and management. However, since the text books are usually American or Australian they impart knowledge about production systems in those countries, which vastly differ from the context here or the needs and demands of the Indian situation. Students are seldom taught about livestock rearing within the Indian context, the climate, the terrain, the people who rear them, the various linkages with our crops and our forests, the different breeds suited for different regions and the environmental concerns of different production systems.

The Indian texts, the body of knowledge which is Ayurveda, different forms of traditional knowledge whether it is our oral traditions or folklore, the social context, the role of women are completely ignored. Students are told that Indian systems are outdated, unscientific and backward. A student emerging as a graduate from these colleges, therefore, when confronted with the harsh realities of the country's production systems is quite unable to deal with it. Even research is propelled along the problems and needs of affluent farmers and is seldom motivated by the genuine problems of the field.

There is another chasm within these universities where veterinary science is the discipline held in awe as compared to agriculture, home science or fisheries. Outside the campus though, this discipline comes under the Ministry of Agriculture. This discrepancy has led to needless tension between the these two disciplines till finally in some states including Tamil Nadu and Maharashtra independent veterinary universities have been formed.

This spells yet another break in the link between agriculture and veterinary sciences because veterinarians are primarily only interested in the clinical aspects of their science rather than production aspects and agriculturists are more interested in cash crops and horticultural crops than food and fodder crops.

Typically, topics like crop and livestock interactions, environment and livestock interactions and fodder security get completely sidelined while more glamorous subjects hold sway.

While upgrading our systems and remodeling our systems, based on the American dream, has become the focus of our policy-makers, farming systems in the west themselves are poised on a dangerous path balanced on a new science called biotechnology.

A Step into the Future ... Modern Livestock Farming Practices

The industry within the livestock sector that has touched dizzying heights in the country is the poultry industry. Traditionally, poultry farming was done in the backyards of millions of rural households. This was the one activity over which women more or less had total control. However, Indian breeds of poultry are poor layers. Moreover, they succumb to different diseases every year and the market for these products is extremely localized. At a larger level, different tiers of middlemen also create difficulties in the markets like all rural markets in the country.

Soon after Independence, a technical cooperation mission, the TCM of the United States of America was invited to begin the process of making dramatic changes in the concept and approach to poultry farming. The country imported around 30,000 White Leghorn and Rhode Island day old chicks in 1959. As the traditional poultry rearers, rural Indian women, were not in a position to manage and maintain these birds, and so professionals stepped in to exploit the opportunity, which ultimately lead to a booming industry.

At present, 75% of the egg, broiler and poultry meat production in the private commercial production units takes place in urban or peri-urban areas particularly around the metro cities of Mumbai, Chennai, Delhi and Kolkata. The states of Andhra Pradesh, Tamil Nadu, Maharashtra, Punjab and Haryana account for 60% of all the egg production. The top five states for broiler production are Andhra Pradesh, Maharashtra, Punjab, Karnataka and West Bengal. Together they produce 61% of the entire broiler output in the country.

Some of the support the industry has received from the government is in the areas of vaccine development and training facilities. The industry is further seeking support for promotion of maize cultivation and increase in both area under crop as well as production per hectare. They are also requesting an effective disease surveillance system and support vaccines to guard against a repeat of the 1994 Infectious Bursal Disease outbreaks, which practically broke the smaller amongst the poultry farmers. Rural poultry production, the commercial way, has been identified as a thrust area for the government.

If we examine this system a little more closely, one realizes that the producers here are not poor Indian farmers, but highly successful private entrepreneurs and professionals.

The system relies on:

- Expensive breeds: pure line breeding of these strains is very expensive. The entrepreneurs are seeking government support for imports.
- Effective vaccination: here again the government supports the industry. Vaccines are being produced and marketed in large doses.
- Expensive feeds: this kind of industry requires quality feed and grain. Under these systems, maize, which can be used for feeding humans would be diverted for feeding poultry birds.
- Medicines: as these animals are very delicate and entrepreneurs are interested in maximizing their profits, these animals are continuously fed hormones, antibiotics and other feed supplements to boost growth. Ultimately, these enter the food systems of the humans who consume them.
- Proper housing management: poultry raised under these systems are raised under deep litter systems and battery cage systems. These intensive systems advocate closed quarters in preference to the free range systems of old-time poultry farmers, because if animals exercise too much they lose weight and the profits are not sufficient.

Livestock rearing has thus become an industry and animals have become factories producing milk, meat and eggs. We now have dairies, piggeries, goateries and rabbiteries producing different specialized products. The returns may be high, but so are the inputs, which sometimes do not even spell out hidden costs, which are borne by others who are less fortunate. Peri-urban dairies operating in the periphery of Mumbai have systematically exploited tribal communities living in the north of

Mumbai for grass. Food meant for human consumption and grown in rural areas is diverted for feeding animals, and ultimately sold in urban markets.

To maximize their profits, farmers will use any means to make the animal produce. Hormones, antibiotics and a range of feed supplements then become essential. Dairy farmers to facilitate milk letdown in cows and buffaloes regularly use oxytocin injections. Oxytocin and similar hormones are expected to make up for fodder imbalances and a deficient diet. The effects of the unlimited use of these compounds are neither well understood nor told to the consumer who purchases these products.

Intensive systems, where animals are kept huddled together cause stress to the animals as well as environmental pollution and also constitute a health hazard because of the build-up of gases like ammonia. Expensive chemicals and disinfectants then have to be used to sanitize these areas.

When competition increases in the market, smaller players in the market also adulterate products to claim better prices. Urea, baking soda, vegetable oils and water are well known adulterants of milk so while India can boast about being the largest producer of milk, the quality of milk remains questionable.

There are a number of reasons why the Indian livestock industry despite all the efforts in the past fifty years really does not stand much of a chance

Box - 7

About Camel Milk

Although amongst traditional camel rearing communities it is taboo to sell camel milk, camel milk finds its way into milk marketing channels more often as an adulterant. Camel milk differs in composition from cattle and buffalo milk. It is quite salty in taste and rather thick. It does not spoil as quickly either. All these factors make it a favourable adulterant. The season for peak yields in camel milk is the hot summer months when milk from conventional sources-cattle and buffalo-are low due to scarcity of fodder and water. Camel, sheep and goat milk are then added to cattle and buffalo milk to keep the supply side high. In Ahmedabad, Gujarat, this kind of milk is frequently referred to as 'Shecago' (Sheep Camel Goat) milk.

in international markets. The reasons could be starting with the poor quality of milk, the unhygienic quality of meat, the existence of numerous diseases in livestock and their products, the unhygienic standards of slaughter houses to the fact that many dairies and cooperatives have failed to deliver. The international resistance to the consumption of buffalo milk has also contributed to our failures in the international market. Realizing this, our policy-makers have had to take another look at policy.

Distorted Visions: Seeing the Poor and their Livelihoods through Opaque Glasses

Vision 2020 is the planning exercise being undertaken by planners and policy-makers for sectoral planning to enable the formulation of future policy. The 'vision' is based on projections and forecasts made at different levels, both nationally and internationally. Delgado's Livestock to 2020, the Next Food Revolution : Food, Agriculture and the Environment (1999), A.P.J. Kalam's *India 2020: A Vision for the New Millennium*((1998), S.Z. Qazim's *National Livestock policy Perspective* (1996), *Swarna Andhra Pradesh Vision 2020* produced by the Andhra Government(2000), *Agro Food processing: Technology Vision 2020* produced by the Department of Science and Technology, TIFAC (1996) are some of the papers which have spelt and predicted different scenarios. The different visions for the livestock sector demonstrate different perspectives. These perspectives include global livestock trade trends, WTO and post- GATT trade negotiations, demographic trends, techno-economic perspectives, institutional frameworks, equity and livelihood issues and environmental and ecological dimensions. An analysis of the projections and future strategies to be implemented shows a lack of vision especially concerning the needs of the poor. The different projections from these papers and the focal point of concern that some of us working with poor livestock farmers feel are presented below:

Projection 1

The overall global vision predicts that with increase in urbanisation, development and larger disposable incomes in the developing countries there is going to be an increased demand for animal products.

***The focal point of concern*:** To meet these demands there are essentially two sources: first, the commercial or industrialized farming systems, which can rapidly cater to the increased demand and secondly, small farmers and producers. If commercialized systems enter, they will displace small farmers completely and wipe out millions of livelihoods. While one hopes that small farmers can improve production to meet this demand, it should not be due to coercive market compulsions and at the expense of their own food and nutritional security.

Projection 2

A number of people in developing countries, including India are likely to have large disposable incomes, which they will then like to expend on animal products like milk, meat and eggs.

***The focal point of concern*:** It is not very easy to predict or project future market preferences and taste. Recent international scares like the Mad Cow disease (BSE) the Foot and Mouth Disease in Europe in the year 2001 and the Salmonella case in poultry may all change and influence taste away from animal products.

Projection 3

Indian livestock produce stands a big disadvantage in the international market. This is primarily because of the quality of its produce. To meet these concerns and give India a comparative global advantage, commercial and industrialized production with strict quality control is the answer.

***The focal point of concern*:** Internationally, there is a growing concern of the indiscriminate spread of commercial and industrialized systems of farming because of their highly negative social and environmental implications. The visions for India completely miss the environmental dangers posed by industrialized systems which are excessive soil and water pollution because of excess nutrients, greenhouse gas emissions and loss of domestic biodiversity, All these factors will ultimately affect the poor who are dependant on natural resources and domesticated livestock diversity.

Projection 4

The turmoil India has been going through since the onset of the new economic policy which includes the elimination of licensing, invitation for foreign investment, measures undertaken under the 73rd constitutional amendment and the minimization of the role of the government are strong indicators of the increasing privatization of various institutes.

It is envisioned that because of their poor performance various public sector dairies and cooperations will be wound up, industrialized farming systems will emerge in a major way and research and education will go out of the public domain and will come under relentless pressure to integrate with the demands of the industry.

The focal point of concern: One can only imagine the inherent dangers of the privatization of essential services and research being mediated by profit-oriented companies where the concerns of the poor (food and fodder security) do not feature at all.

Projection 5

Present demographic trends as an indicator for steering growth. Commercial poultry, buffalo rearing, cross-bred dairy cattle rearing and pig farming are seen here as sectors to be promoted.

***The focal point of concern*:** Mono-gastric species like the pig and poultry are being promoted in the west as they are supposed to be more efficient utilisers of feed. However, the requirements and needs of people in our country and their patterns of resource use are quite different from developed countries. Current demographic trends (See Fig 4.2) in fact show a very rapid increase in four sectors with regard to local cattle, goats, indigenous poultry and commercial poultry. Local cattle, goats and indigenous poultry are the animals reared by the poor. As mentioned earlier in this chapter, when pressures of the environment and natural resources begin to squeeze the poor, they rear animals that are less demanding and which can be traded quickly. This in essence explains the massive spurt in the population of these animals despite the fact that there has been no incentive from the government. The commercial poultry sector on the other hand shows an

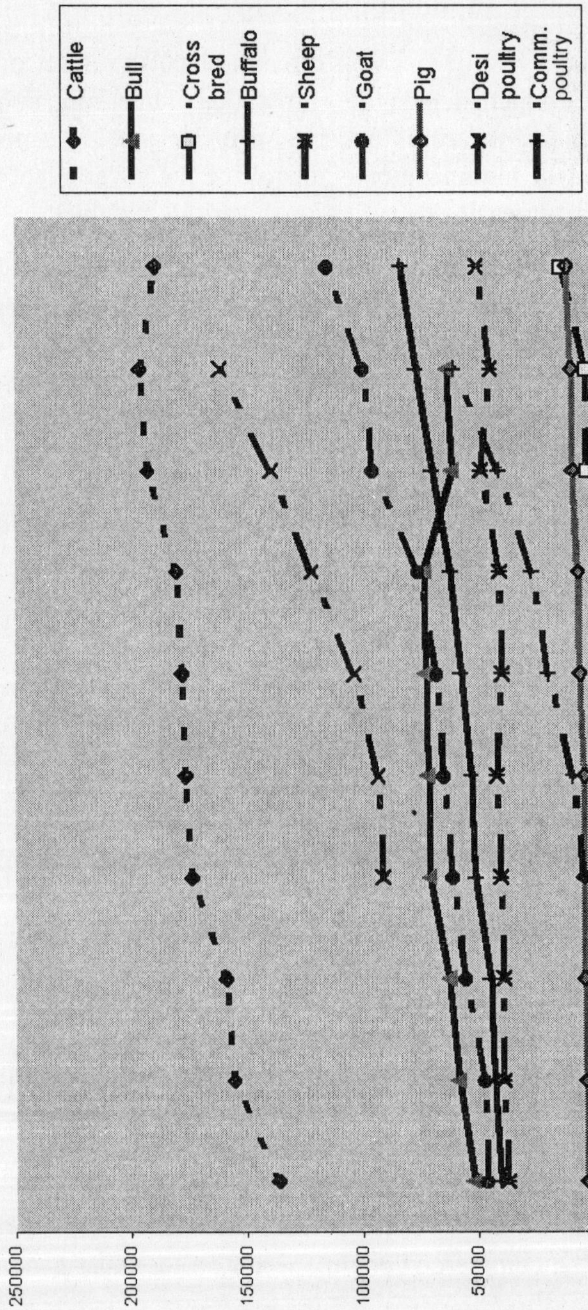

Fig 4.2 Livestock demograpic trends

enormous spurt of growth because as explained earlier, it has received extraordinary support from the Government.

What is alarming though is the negative growth rate of work animals. If the poor cannot afford mechanization on their farms and they do not have good quality work animals, how will they farm?

Projection 6

With a decrease in livestock numbers achieved by the introduction of high-producing animals, there would be less pressure on grazing and feed resources.

***The focal point of concern*:** The more prolific animals require far more food and of far better quality than indigenous livestock. Where is the land or where can one find the resources to grow this type of feed? Can we afford to divert our agricultural land to feed animals at the cost of feeding our people?

Projection 7

Modernization will do away with the environmental dangers posed by outdated slaughter houses, the leather and tanning industries.

***The focal point of concern*:** Only the rich can afford the benefits of modernization. Where then are the employment opportunities for the poor? Traditionally in India, the poorest castes have been employed in jobs such as leather processing and tanning. Do we have alternative visions for them?

Projection 8

Dairy farming is expected to go the way commercial poultry farming has gone in India and it would give us the global competitive advantage. But it will spell untold misery for millions of resource poor households.

***The focal point of concern*:** Unfortunately, none of these documents have any visions or answers for the misery of the poor nor do the projections or visions inquire into the livelihoods of small farmers who rear species other than dairy cattle and buffalo such as goats, pigs, sheep, yak and poultry in their backyard.

The Future of Livestock ... A Jump Ahead

Indian livestock industrialists and their promoters are still comparatively very small players in the field. Standing on the edge of a global economy, the equations are very different. Agriculture is a major subject in trade negotiation with the West and countries like the U.S. are making a strong appeal for the elimination of protection for small farmers in Europe and Japan. Where does this leave our farmers? After all even our 'large' farmers barely make the 'small' grade when compared with farmers of the West. David C. Korten (1998) describes the fate of small farmers in the U.S.:

> From 1935–1989, the number of small farms in the United States declined from 6.8 million to under 2.1 million. During the same period the U.S. population roughly doubled. As farmers have gone out of business so too have the local suppliers, implement dealers and other small businesses that once supported them. Entire rural communities have disappeared. Meanwhile, the major U.S. agribusiness corporations have grown and consolidated their power. The top ten farms in the United States are now international agribusiness corporations with names like Tyson Foods, Con Agra, Gold Kist, Continental Grain, Perdue Farms, Pilgrims Pride and Cargill—each with annual farm products sales ranging from dollars 310 million to 1.7 billion.
>
> Two-grain companies, Cargill and Con Agra, control 50% of U.S. grain export. Three companies—Iowa Beef Processors (IBF), Cargill and Con Agra—slaughter nearly 80% U.S. beef. One company—Campbell—controls nearly 70% of the U.S. Soup market. Four companies—Kellogg, General Mills, Phillip Morris and Quaker Oats—control nearby 85% of the U.S. cold cereal market. Four companies Con Agra ADM Milling, Cargill, and Pillsbury- mill nearly 60% of U.S. flour. This concentration is in part the consequence of 4100 food industry mergers and leveraged buyouts in the U.S. between 1982 and 1990—and the consolidation process continues.

He goes on to say that, '...the farmer receives a lower price and the consumer pays a higher price than either would have obtained under conditions of true competition. This is the system that the dominant agribusiness companies hope to extend to the world'.

India is now part of the global economy and therefore the international markets already have a presence here. The wool produced by Indian sheep finds no market in an otherwise large woolen garment industry in the country, as wool from South America and Spain are cheaper and finer. Indian supermarkets are already stocking meat and milk products from different countries at competitive prices to Indian producers. Most industrial nations are already over-producing milk. Thousands of small dairy farmers in Europe and North America today stand threatened by the glut in the market. It is expected that milk from Australia and New Zealand will flood the international as well as the Indian markets and it will take all the energies of the dairy industry to keep the prices on par with prices suitable to Indian producers. Where then do our poor farmers stand in the face of such global giants?

Technology which Terminates

Soon research in universities will be financed and propelled by these giant corporations. There are already many attempts in this direction. In most developed countries, commercial animal farms rely on single breeds. In fact, almost all the pigs reared under commercial farming systems in Europe and North America correspond to the same breed. Also 90% of all North American dairy cattle and 60% of all European cattle belong to one single breed the Holstein. Further, it is estimated that by 2015, the genetic diversity within this breed will correspond to that of only 66 animals. Organised poultry farming across the world relies on a few multinational companies who have developed a handful of breeds for their supply of stock. To get new stock they have to be purchased from breeding stock, which are controlled by these corporate giant companies. This has ensured that genetic diversity is all but wiped out.

Rifkin(1995) points out that the global agribusiness complex hopes to extend its arms to gene based agriculture and animal husbandry in this century. In Australia, scientists have succeeded in producing genetically engineered pigs that are 30 % more efficient and brought to market seven weeks earlier than normal pigs. Farmers who do not have access to genetically engineered pigs get wiped out.

To increase productivity, researchers are currently experimenting with insertion of a growth hormone gene directly into the biological code of the animal at the embryonic stage of development so that the adult animal will produce more milk without needing injections. With more milk being produced in an already saturated market, enormous price supports and subsidies will have to be given by the Government to keep dairy farmers in business. Companies like Monsanto, American Cyanamid, Elli Lily and Upjohn have spent more than a billion dollars in research and development to bring Bovine Growth Hormone or Bovine Somato Trophin to the market. They subsequently approach governments for subsidies to introduce these drugs in markets of developing countries. The potential harm these drugs can induce is seldom made public. Frontier techniques like genetic engineering, gene splicing and cell fusion techniques allow scientists to cross all biological boundaries by recombining genes from totally unrelated species. As scientists reduce life and living organisms, microorganisms, plants and animals to their constituent building blocks, they can actually begin to recreate new life forms and organize life as a manufactured process.

Apart from this, technology and machines are also being created to make work on farms more 'efficient'. Animal husbandry management in developed countries already use many sophisticated systems. As labour on farms becomes expensive, robots are being developed who can replace human farm hands in animal farms. Already milking machines have all but completely replaced human milkers in developed and commercial farms. It is soon expected that machines and robots will replace professional sheep shearers too.

Computerized systems are already in use to feed pigs and dairy cows. Each animal wears an identifying pendant into which the farmer codes the ration number of the animals and the amount of grain to be fed each day.

At the feeding station special sensors identify the pendant. This releases the quantity of ration the animal is programmed to have. Similarly, pigs enter a maze individually which eventually leads to a trough where concentrate rations for the day are delivered. The time each pig will take is also carefully computed and programmed so that time is not wasted.

And perhaps there will come a time, when there will be no farmers, no helpers, no women tending to calves, no pastoralists, no stockmen, no horsemen, not even veterinarians as computers and robots will diagnose and dispense medicines. The human race would have replaced itself with clones, computers and robots.

5

It Must Not End: Towards an Alternative Policy

We hope that the concluding paragraph of the previous chapter is not our destiny and that the powers of reason will steer us clear of a singular path of destroying all that is valuable. We hope that our history and our traditions will give us the stimulus and the opportunity to explore different ways.

The enormous diversity in our country and our culture are in fact the main keys, which can prevent humans from becoming automated robots and animals or becoming production factories. For all those who believe that being pro-environment is being anti- development, the converse in fact is true. Preserving the environment and the diversity within would lead us on a development path which is more humane, ultimately more sustainable and viable.

Exploring Diverse Ways ...

When we talk of diversity in the context of livestock rearing, we do not merely talk of diversity in livestock species, breeds and produce but have to extend the idea to diversity of resources, patterns of rearing, and other aspects which have a direct or indirect bearing on the welfare of animals.

Diversity of livestock species: mixed farming systems

There is enormous sense in mixed or multi-stocking.

- The different animals do not compete for the same resources in a particular area. For instance in an arid area if sheep and camels are raised together, the flock of sheep should be made to graze on the grass while the camels consume tree leaves. This prevents overgrazing of a particular type of fodder.
- By grazing different species on the same area, the life cycle of parasites gets broken. Many parasites of animals are host-specific and when they get into the system of a new host their life cycle gets broken. For example,

many species of worms are specific to cattle. By grazing other animals like sheep also on the same pasture, these other animals which are not their natural hosts, ingest the eggs of these parasites, and consequently the eggs cannot multiply.

- Mixed farming also prevents the spread of epidemics. First, the spread is delayed, as there are fewer animals of one species within a farm. Secondly, should an epidemic consume one species, there would always be another species, which the farmer can rely on. Every year epidemics claim almost 80% of the poultry amongst backyard poultry farmers. If these farmers had been dependant on poultry alone, in other words literally, 'put all their eggs in one basket', they would not have been able to recover from the crisis. However, since many of them also keep a few goats, sheep or cattle these then help them tide over a difficult period.
- Different species also provide diversity of produces such as meat, milk, wool, and dung as the case may be for home consumption and for the market. Diversity prevents oversupply of one product and also helps staggering and spreading income availability through out the year, as many of the products are seasonal.

Diversity of breeds

Within a species different breeds have different strengths. In cattle, we have draught as well as dairy breeds. Within draught breeds, there are those which are good for the plains and others which are good for hilly areas, some are good for hot and dry climates and then there are others which are good for heavy rainfall areas. Amongst dairy breeds there are those which mature early, those which give a high percentage of fat in milk, others which have a good feed conversion ratio, yet others which can withstand drought and nutritional stresses and even those which are resistant to diseases. In regard to sheep, there are wool breeds, meat breeds and even breeds that provide milk.

This genetic variety provides for a large gene pool, which can then be used for developing new strains and varieties depending on the need. This vast reservoir or gene pool provides for an uncertain future, which may make new demands. This gene pool, however, is in danger of being lost if there is indiscriminate promotion of one or two breeds in the manner similar to what is being done with industrialised pig and poultry farms in developed countries, commercialised poultry farming systems within our country or in the case of dairy cattle, for example, by promoting chiefly Jersey and Holstein Friesian breeds. Such methods lead to the dilution or the loss of a vast and diverse gene pool which it may not be possible to recover at a future date.

Diversity of resources

Typically, fodder for animals comes from three sources: farmlands, grasslands and forests. By preserving all three we ensure that there is always one source, which can provide for animal food. Buying and transporting fodder are always expensive options so it is very important that requirements of fodder should be met locally as much as possible. For this our fodder resource base have to grow enough for local consumption and create a surplus whenever possible.

Natural calamities, droughts and disease are the three major factors, which can unexpectedly affect fodder availability. However, it is highly unlikely that all three occur together or affect all three sources at the same time.

Even in major catastrophes like earthquakes and cyclones, the severity is often localized and if appropriate precautions and plans are taken in advance to ensure that our farmlands, grasslands and forests are not degraded, fodder resources from areas which are not severely affected can always be transported to the worst affected areas. Or sometimes it may just be more feasible to transport animals to areas where surplus fodder is available. For instance, trucks to fodder-rich areas in Punjab often transport sheep from fodder-scarce areas in Rajasthan.

Ensuring Farm Diversity

Rather than promoting mono-cropping with a single cereal crop or cash crop it is advisable to crop a variety. Not only does it provide a wider nutritional base, but as with the case of animals, should one crop succumb to disease or pest, there would be another to tide over a bad patch. Also, intensive cropping with the same crop year after year can deplete the soil of different valuable nutrients.

Crop diversity must include a mixture of:

- *Cereal crops*: the grains of which can be consumed by humans and the stalks of which are good fodder.
- *Millets*: Especially for landholdings in fragile areas, on hill sides and marginal lands for most millets such as sorghum, pearl millet and fox tail millet are hardy, can tolerate water stress and are extremely nutritious both as human food and fodder.
- *Pulse crops and legumes*: For these are an important protein source for human and animals and also help fix nitrogen in the soil.
- *Oil seeds*: For providing the fat component in the diet of animals and humans.
- *Vegetables and fruit*: Many vegetables and fruits have both nutritional and medicinal value.
- *Weeds:* Even weeds that grow between plants are useful as animal feed. Many weeds are also important as medicines. The indiscriminate application of herbicides destroys an easily available and cheap source of fodder as well as locally available medicine.

By strengthening our farming systems, we can ensure that fodder requirements of landed farmers are met as much as possible from their lands so that they do not compete with the landless for common grazing resources. While there are no easy solutions to the conflicts on common grazing lands, at least, some pressure on them can be eased by enhancing fodder availability from farmlands.

Maintaining the Diversity of Grasslands

While many of our grasslands have been reduced to the status of wastelands or barren lands due to neglect, there have been attempts to restore our grasslands especially by the forest department and watershed programmes. Unfortunately, most of the grasses recommended for planting are limited to a few species, like *Stylo hemata* or *Cenchrus ciliaris*. Some of these are exotic or from other agro-ecological zones. In some cases, trees and shrubs have been planted on grasslands thereby reducing the grass cover.

We have to consider the fact that our grasses have to provide nutrition not only for domestic species, but also for our wild fauna species. Many of our traditional grasses are probably the best species for their area and the indiscriminate introduction of an exotic may create future imbalances. Instead, an attempt has to be made to systematically research the initial grass cover and restore it in a phased and organized way to its original state.

There have been various attempts to restore grasslands by preventing grazing and allowing them to regenerate. While this has worked with reasonably good results in many watershed areas, one area of concern is that if the grass is not cut or grazed it leads to fire hazards. Further, by preventing animals from entering, the natural fertilization of grasslands and dispersal of seeds does not happen.

Studies therefore need to be conducted into carrying capacity and rotational grazing methods wherein these lands can be managed better. It is also critical to ensure that the major stakeholders or, in other words, the livestock owners participate completely in the process so that fodder resources are equitably shared.

Preserving Diversity in Forests

As a resource, the wealth from forests is unlimited. Recognizing this fact, many countries like Norway, Canada, Austria and Switzerland are extremely careful not to destroy their forest cover. The forests are also the home of a number of communities as well as of wildlife. Conflicts consequently arise when competition for resources increase and non-forest dwellers also begin to stake their claim on forest produce. So before any programme on forests is undertaken it is critical that the different biological interactions and dependencies are understood and studied. It is also essential that the traditional rights of forest-dependant communities are well understood and their needs taken into consideration.

Within forests, care must be taken to ensure that single species and mono-cultures are not randomly promoted or planted. Different varieties of trees, depending on the type of forest need to be planted. Further, preference should be given to trees, which have multi-purpose value in terms of food, medicine, timber, fuel wood etc. These different trees are also important for maintaining local biological systems. Many creepers, vines and grasses grow in the shade of or around specific host plants. Many of these creepers and vines have considerable food and medicinal value. By eliminating the dominant species we also eliminate species, which are dependant on them. A very simple example to illustrate this is the creeper, *Tinospora cordifolia*, which is well known in Ayurveda as a medicinal plant. This creeper grows on the *Neem* tree. If we were to accidentally remove all *Neem* trees from an environment, we would also lose *Tinospora cordifolia*. A detailed understanding of biological systems, therefore, becomes very important and needs research.

Sustaining Livelihoods

Diversity of systems

It is quite evident now that grasslands and forests especially in marginal areas cannot be converted to farms. Conversely farmlands cannot be converted to pastures and forests. Nor should they be. Recognizing this limitation, before we promote or reject a system, a more thorough study of the system needs to be initiated exploring them from different dimensions. New livestock development programmes need to be developed regionally taking into consideration different ecological niches and local needs. Blanket criticism of traditional systems should also be avoided till they are better understood. One of the cardinal mistakes of livestock policies in the past has been the attempt to push uniform policies throughout the country.

Migratory groups inhabit semi-arid and fragile regions. If we were to force them to settle down and should there be a drought the next year, it would be impossible to ensure that sufficient quantities of fodder are transported to every herder in every village in a fair, equitable and just manner. We only have to look into our Public Distribution System for food to realize some of the problems within. On the other hand, herders have the option of reducing the pressure on fragile areas by exercising the option of moving out. They can also exercise the option of reducing their flock.

There are fine examples of herding systems as well. Farmers in the Alps continue to shift their herds seasonally. In the winter, animals are brought to the valley and housed there, while in the summer they are up in the Alpine pastures. Great care is taken to preserve the tree cover of the crest of the hills so that the natural pasture on the slopes is not destroyed. Different pastures are rich in different herbs and are supposed to produce different qualities of milk and consequently different varieties of cheese. This knowledge has been harnessed by farmers to their advantage wherein these farmers, even when they are herding their animals in remote pastures, are able to convert their surplus milk to quality cheese of many different types. These varieties of Alpine cheese are internationally famous and command a good price because of the unique (region-specific) conditions, which go into producing them.

On the other hand, it may not be such a wise idea to promote large scale livestock development programmes, which involve expensive animals and open grazing methods within heavily forested areas because there are dangers of predators and of conflicts of grazing areas with other wild herbivores. A market-dependant livestock system will also not work here, as access to markets would be difficult. In such areas, it may be preferable to consider animals, which can adapt easily to stall feeding like buffaloes, backyard poultry or pigs, which can then be used either for home consumption or for local markets.

Likewise in areas that are fragile, degraded or where high quality fodder is not possible, expensive type of livestock rearing or animals that demand high quality fodder should not be promoted. Instead breeds and varieties, which can easily adapt to fodder stress, may be considered.

Diversity of produce

One has already seen how present trends in the global economy are driving small farmers out of their business. Crop farmers in irrigated parts of the country are being forced to change their crop every year to keep their heads over water. From traditional crops like jowar they have had to shift to wheat or rice and then to sugar cane followed by cotton, grapes, figs, citrus fruit, tomato all in an effort to stay ahead of the game, but constantly governed by the dictates of a global market. The problem here is that these trends are then imitated by other farmers and there is excess in supply leading to a fall in prices. There is always considerable risk when one caters only to a distant market. First, there is almost nothing one can do to control it. Secondly, the production is sometimes at the cost of a necessary item for home consumption or local consumption. Diverse produce, on the other hand, cater to different needs including the household, different customers and different markets.

This keeps different options open and allows farmers greater flexibility. It would, therefore, seem better to take stock of our resources and develop a range of products traditional and new, based on local specialities.

Finally, markets continue to be controlled by middlemen. Direct and cooperative forms of marketing will have to be explored and the capacity of farmers will have to be strengthened to ensure that the stranglehold of these middlemen is broken.

6

Reorienting Ourselves

A change in existing systems would require reorienting the way we look at animals, their herders and the healers. It would also require reorienting the programmes designed for them, the interventions planned for them and the way the policies have been formulated.

Reorienting Programmes

Planning with people

When we talk about diversity of systems, it also becomes important to consider the preferences of people living in an area. While a common criticism of our farmers is that they over-stock poor quality animals, and should instead keep few good quality animals, our livestock development schemes are often guilty of promoting the same. Schemes are designed where landless farmers can avail of credit for keeping units of 10 cross-bred dairy cows or 50 goats. Although great care is apparently taken to calculate every last detail they often miss the one major component, which is feed. Landless farmers cannot grow fodder and are therefore dependant on buying fodder if their unit is to succeed. If for some reason the production does not match up to their expectations due to illness, accident, death, which is quite often, the units become non-viable. Schemes are also designed for women from poor households, where the different inputs of fodder, labour and veterinary care are not in place. Market linkages are also weak.

Unfortunately, financial institutions only give credit for schemes designed by them and seldom listen to what people want. Farmers often want very little credit for purchasing one animal or a few chicken. Farmers are also more comfortable keeping animals they are familiar with rather than stocking a species they have never reared before. Credit and financial support must therefore be given to beneficiaries only after taking note of the needs of the people.

New programmes that are initiated need to be backed up by strong training support as well as through close surveillance and monitoring. This too is possible only when there is support from the people who are benefiting from the programme.

An example from the East Godavari district is cited below to show an example of planning programmes with people.

Box - 8

Planning with People

For thousands of tribes living in the forests of India, poultry is one of the most prized assets. Besides cock fighting, a popular sport, poultry is also needed for paying bride price, for sacrifices, at festivals and for offering at the table to favoured guests especially sons-in law.

The Hindi word *Asli* translates as 'the real thing'. A good *Aseel* can fetch anywhere from between Rs. 500 – Rs 1500 in the market. There are a number of strains or variations of these birds recognized by the people of this region. *Aseel* birds are not prolific egg layers with the average number laid per year being 36 – 60 (in comparison with western exotics like white leghorn which lay over 200 eggs /year). However, since they are primarily kept for meat, the egg laying capacity does not really matter. What matters though is that every year almost entire flocks get wiped out due to disease either bacillary white diarrhea or Ranikhet. This loss is becoming difficult to replace every year.

Realising these programmes while working in the areas of East Godavari district three organisations, Girijan Deepika, a tribal peoples organization working on different issues concerning their region, ANTHRA* and YAKSI§ did a participatory appraisal with poultry owners in the region and tried to assess the major reasons for the losses in poultry.

They found that poor nutrition, loss to predators, and disease can almost cost a family two thousand rupees a year.

(contd.)

* An organisation working on livestock development

§ A development organisation working with tribal groups

(Box - 8 contd.)

This loss was also threatening the gene pool of the *Aseel*. On the other hand government poultry programmes were promoting hybrid varieties like the *Giriraja*, which if introduced indiscriminately into the environment would prove a major threat to the *Aseel* gene base in the region. The results of the appraisal were shared with the community of poultry owners. Women who are the main caretakers of poultry in the household came forward in an initiative to protect and preserve their birds. The programme had different features:

- The training of village level health workers who could administer preventives (vaccination and deworming) as well as a few medicines if animals fell sick.
- Teaching of improved management practices to the village community including the importance of providing birds with clean water, improved feeding by encouraging the community to grow traditional crops: millets, pulses and oil seed thereby providing the birds a wider nutritional base. Keeping the poultry house clean. Regular fumigation with different leaves to get rid of external parasites.

It has been found that the mortality rates for birds have decreased enormously. Production levels have also increased almost two fold.

Once the mortality rates went down, the women re-initiated a traditional system of sharing called Vaata to build up gene pools. Girijan Deepika got together a pool of good birds. Following which they identified villages with women's groups (called *Gotti* in the region) who were willing to collaborate in the programme. While the village received a male bird, women beneficiaries identified by the *Gotti* members received *Aseel* hens. Once the birds multiplied half their brood had to be returned by the women beneficiaries to the *Gotti*. These birds formed a pool, which would then again be distributed to a new set of beneficiaries in the village. Preference is been given to women who are asset less, single or landless. When all the women of the community have benefited from the programme, the Gotti then returns half their assets to Girijan Deepika who houses them for approximately three months. Those showing desirable traits are distributed to fresh *Gottis*, while those that do not match up to desired performance levels are sold in the market.

(contd.)

(Box - 8 contd.)

Each person benefiting from the programme is encouraged to maintain records of their brood.

The Girijan Deepika model illustrates different aspects of participatory planning.

a. Working with people to identify their genuine needs and problems.
b. Involving people in the planning process.
c. Respecting local knowledge and strengthening it to achieve the desired objectives.

All of these are extremely essential when new programmes are planned.

The programme is also managing to conserve an indigenous breed, promote local diversity of agriculture, improve animal health, build up a cadre of local village para- professionals as well as animal health workers. All these in turn should positively lead to better economics and better livelihoods in the area.

Reorienting policy

Thrust areas

Preventive medicine

Every year a large number of our animals succumb to disease. Foot and mouth disease in cattle, sheep pox in sheep, Ranikhet and Salmonella in fowl claim large numbers of productive animals. Yet, there has been little effort made by the state to strengthen preventive medicine. There are three major lacunae in the effective delivery of preventive medicines.

1. *Lack of suitable vaccines*: For a number of diseases sufficient quantities of vaccine are not available. In the case of emerging diseases like blue tongue there are no vaccines available in the country. And with some diseases like foot and mouth disease and sheep pox, the strains of virus keep mutating and the production of vaccine does not keep pace with the mutations.

2. *Lack of infrastructure*: Most vaccines need to be transported on ice. This in itself is difficult in a tropical country, but becomes more so in remote areas where electricity and ice are not available. While there

have been attempts to bring out thermostable vaccines, these are not easily available. In fact, a thermostable vaccine for Ranikhet (New castle) disease has been developed in Australia and the technology, in fact, has been transferred to a number of African countries, but for some reason this vaccine is not being made easily available in the country.

Poultry vaccines are packaged in large packages of 100 doses or more, which is useful for large poultry owners but quite useless for small farmers practising backyard poultry who lose a large portion of their birds to disease every year. By neglecting this population, we are neglecting breed diversity and product diversity.

3. *Lack of personnel to vaccinate*: Although the national perspective policy paper suggests that there are far too many veterinarians graduating every year, the truth is most of them want to be located near urban centers where facilities are available. In the remote areas, the nearest veterinary centre is often 25 or more Kms away and farmers find it difficult to take their animals there. It also becomes difficult for the person responsible in that centre to cater to the demands of such a wide area. Instead properly trained animal health workers from each village can manage the immediate problems including regular immunization and other preventives. These programmes have been successful in a number of areas and definitely need more consideration.

Box - 9

The CHIP Project

The Camel Husbandry Improvement Project (**CHIP**), is a joint endeavor of the School of Desert Sciences in Jodhpur and the League for Pastoral Peoples, Germany. The area of operation is the Pali district of Rajasthan and the project has been operational in the area and making adequate efforts towards the promotion of camel welfare since 1996. Initially, the project undertook an action research project to determine the major problems faced by camel rearers in the area. Some of the major problems identified were the absence of good breeding bulls in the area, a high incidence of mange, trypanosomiasis, abortions probably related

(contd.)

(Box - 9 contd.)

to trypanosomiasis and insufficient economic returns from the activity of camel rearing due to changes in existing practices. The League for Pastoral Peoples along with a local NGO, Lokhit Pashu Palan Sanstha (LPPS) has intervened in this sector by obtaining quality bulls and making them available to the Raika community.

They have also tried to make some interventions in the health sector by involving local veterinarians as well as camel breeders. Posters and educational material on the prevention and treatment of these diseases are available at the LPPS office in Rajasthan. LPPS is also keen to initiate a programme for the training of para-vets who will be trained to use as much of traditional medicine as possible.

Another intervention made by them is the better promotion and marketing of camel milk as an additional source of income for camel breeders. LPPS is trying to achieve this by providing the necessary linkages between camel breeders and consumers, which are tea stall owners.

Perhaps the greatest contribution made by the project though is the creation of awareness in regard to the issue of the status of camels and camel rearing in the country.

Alternative medicine

Another area where a lot of work needs to be initiated is in alternative medicine for animals. While on the one hand, proper veterinary care is not available at the village level because of which a large number of animals fall sick or die, on the other hand traditional medicines including turmeric, neem, fennel and asoefetida are known to cure a range of simple complaints in animals like diarrhoea, bloat, wounds. By ignoring these medicines and their value we are simply missing out on an excellent alternative to expensive medicine.

We are fortunate in having codified texts like the Ayurveda, and Unani that are still practised. Efforts need to be directed towards integrating the different systems of medicine still in practice in the country.

Box - 10

The Kurubas of Medleri

When the gods created this world, there were seven sheep. These sheep were looked after by each of the gods in turn. The last turn fell on Parmeshwara who being busy with other activities could not look after them. The fates conspired and these sheep were consigned to the nether world where they were to be burnt in an oven. At the same time, Undal Padmanna, a kind man took birth and Parmeshwara gave him the responsibility of rescuing these sheep and looking after them. Undal hurried to rescue his charges, some of whom were already dead. Those that remained were burnt black. Undal tended them, nursed them back to health and vowed to look after them forever and thus it came to be that the descendants of Undal Padmanna, the Kurubas, were entrusted with the responsibility of looking after the black sheep of the Deccan.

The Kurubas are a pastoral community who rear sheep under semi-migratory systems in the Deccan regions of Karnataka. Similar groups called the Kurumas and Dhangars exist in Andhra Pradesh and Maharashtra respectively.

The Kurubas earn their living through the sale of lambs, dung and wool. Every year at the end of the rainy season and the beginning of winter, the Kurubas begin their annual migration westwards towards the uplands or Malnad in search of fodder for their animals. They fertilize the fields of farmers and are offered payment in return. They only return to their home villages with their sheep when the rains fall again after the summer is over. The Kuruba women spin wool which is then woven into blankets used by the people of the area.

Like other pastoral groups, the Kurubas too are on the crossroads with the winds of development forcing them to change their pattern of living. However, a unique project in the region gave the entire effort a new direction. There are very few genuine development interventions with pastoral groups in India and the IDS interventions is perhaps one of the best that one has come across in terms of overall impact.

India Development Service (IDS), an NGO in the Dharwad district of Karnataka became conscious of the problems faced by shepherds in 1982

(contd.)

(Box - 10 contd.)

when a large number of sheep succumbed to Rinderpest, a fatal disease of ruminants.

After the 1982 epidemic, IDS mobilized local animal husbandry and veterinary officials to determine ways by which such an epidemic could be prevented in the future. Curative and preventive veterinary services were identified as the most urgent interventions needed. This lead to the initiation of the Sheep Health care programme where shepherds were trained as Village Veterinary workers (VVW). IDS soon realized that only technical interventions were not sufficient and community interventions were also necessary. At this stage, they began organising shepherds into *sanghas*[2] or groups.

The major interventions of this programme were as follows:

The Sheep health programme

Training of the Village Veterinary Workers (VVW)

A few shepherds were selected for training as Village Veterinary Workers. The training was conducted with the help of the local Animal husbandry officials of the Sheep and Wool Development Board and the Chellikere Sheep Breeding Farm. The initial training was for a period of 21 days and was followed up with weekly meetings for a year followed by fortnightly meetings for six months.

A lot of emphasis was placed on preventive medicine, deworming and vaccinations. Simultaneously, the Animal husbandry Department was persuaded to hold vaccination and deworming camps when the shepherds were on migration. Shepherds were also taken on exposure trips to formal farms in order to enable them to see the different aspects of sheep breeding.

Gradually, IDS withdrew monetary support to their VVWs. However, the communities came forward and decided to pay for the upkeep of the VVW. This is, therefore, an example of an Animal Health Worker Programme which has proved sustainable. A few *Sanghas* decided to pay the AHW a monthly stipend , some even provided their Village Veterinary Worker with a motorcycle. Yet another made the Village Veterinary Worker a son in-law of the village and provided him with a home.!!!!

(contd.)

[2] Kannada word for groups

(Box - 10 contd.)

Sheep breeding interventions

In 1984, on the advice of the local AHD, IDS began a programme for cross breeding the local sheep with exotic breeds like the *Correidale*. Later, *Rambouillet* sheep were also introduced. However, this scheme was not a success as the crossbreds showed high mortality and morbidity during migration. IDS therefore discontinued the programme.

Interventions with fodder development

IDS realized that another area which would have to be addressed was fodder availability. Initially, the plan was to establish a community fodder farm. However, this did not work. Later the idea of regenerating wastelands was evolved and shepherds participated actively in raising seedlings as well as broadcasting grass seeds. This was totally in contrast to the normal perception that pastoral communities only consume and destroy grasslands. The IDS experiment proved that with correct motivation, shepherds were more than willing to participate in programmes to regenerate the environment they and their animals depended upon.

Community interventions

The formation of shepherd sanghas

IDS realized that for the shepherding community to make their demands heard they would have to collectivize and form village level sanghas. These sanghas then got together and formed Block level Federations. Each sangha then elected 1–2 members to represent them at the federation level. All sangha members were expected to make contributions to the federation.

Through these Sanghas, the shepherds found that they could collectively discuss their problems and find ways to solve them. They could collectively organise the purchase of medicines and the purchase of vaccines which are often not available in individual doses, but have to be purchased in bulk. They could also get political representation in the local village panchayat. Earlier they were considered a backward group, but now their demands were heard. Through the sangha, shepherds were also able to negotiate grazing rights and make the government aware of their problems.

The IDS experience goes to show that there can be programmes which are designed differently and address a range of issues. It also goes to show that rather than push programmes designed at the top it is better to work with people and design small interventions at different levels.

Reorienting research

Scientists in the west have already become aware of the ill effects of the misuse and the overuse of synthetic and chemical compounds including antibiotics. Research is being initiated on a variety of new compounds including pro and pre-biotic agents as well as a large number of traditional Indian spices to enhance production. The west is also growing increasingly aware of the different medicinal plants and the different systems of medicine in their search for new and alternate medicines. Indian breeds of animals have often excited interest abroad. For instance, the twinning gene in sheep is often traced back to the *Garole* breed of Bengal. By delaying researching into these issues, we are only making it easier for somebody else to stake their claim on the ownership of such research.

Research therefore needs to be undertaken on Indian breeds of livestock, their strengths and suitability for different regions. Research needs to be undertaken into the properties and uses of our medicinal and fodder plants and their appropriateness. Research also needs to be initiated into the effects of different interventions on different groups of livestock owners. There is also a need to research and develop new products, as well as create new and different markets.

Most importantly research needs to reorient itself to the genuine needs of livestock owners especially poor livestock owners in the country and attempt to address their problems rather than be directed by market forces and private interests.

Box - 11

The (IKAH) Project of ANTHRA

ANTHRA is an NGO working with poor rural households helping them develop programmes by which livestock rearing activities can be strengthened. Since 1996, ANTHRA has been involved in a unique project better known as **IKAH** or **Indigenous Knowledge in Animal Health** that combines research, small pilot experiments and training in an effort to mimimise losses and supplement livelihoods. The project

(contd.)

(Box - 11 contd.)

is operational in six districts of India in distinct agro-ecological zones in the states of Maharashtra and Andhra Pradesh. Different community-based organizations including tribal peoples organizations are collaborators of the project. The project has been sponsored by SDC/IC (Swiss Agency for Development Cooperation, Inter-Cooperation). The different aspects of the project are:

Research

Documentation of traditional livestock rearing practices: This has been important in the effort to understand the context, the local constraints as well as the strengths and capabilities that can be built on. What is unique about the project though is that much of the documentation has been done by barefoot field researchers, trained by ANTHRA, who have collected data from their villages. This has ensured that the local communities are fully involved in the research process and have participated in the development process from the beginning. Through this research, ANTHRA has documented over 700 different traditional remedies used for treating approximately 70 different conditions affecting domestic farm animals and 14 kinds of conditions affecting poultry. Over 600 plants, both medicinal and fodder, traditional practices of feeding, grazing, watering, storing of fodder and grains, economics of livestock rearing in poor farming households, traditional housing patterns for local breeds and the role of women in livestock rearing have been carefully studied and documented in detail. Special databases have been created and developed and the information stored in computers. The most promising practices, which have been selected by the communities themselves in regard to their traditional knowledge, are being validated by the different communities through a research protocol specifically developed for the process. Experts in the field of veterinary science, ayurveda, botany, systems analysis, pharmacology and the social sciences have been members of the Technical Committee along with farmers, livestock herders and development practitioners. Extensive directories of the plants along with their properties, herbaria of the plants and directories of healers who are well known in their regions for treating animals using traditional medicine have also been created. The information is then consolidated, analyzed and then returned to the communities in the form of books, newsletters, and other relevant educational material.

(contd.)

(Box - 11 contd.)

Training of animal health workers

Through their work, ANTHRA found that it was necessary to combine elements from both modern and traditional methods to ensure that the most effective, sustainable and cost effective practices reach the poorest farmer. For this it was necessary to have trained Animal Health Workers from the local communities who could ensure the delivery of these services. Many of the barefoot researchers have been trained as barefoot veterinarians. Many women have been included in the training programme. This has been absolutely essential as women are responsible for anywhere from 60%–90% of the daily activities involved in livestock care and many of the activities like cleaning of sheds and feeding which directly impinge on the health of the animal are performed by women. The training encompassed preventives including deworming, immunization, improved management practices and the use of validated local remedies for treating simple cases afflicting animals. For more serious cases, the Animal health workers are encouraged to contact more experienced people and professionals. Today, these animal health workers have been responsible for dramatically bringing down the mortality and morbidity of the animals in their villages. Additionally, they have formed state level networks through which they are able to deal with different issues they are faced with such as lack of resources, both in terms of finances as well as medicines.

Pilot programmes

Conservation of biodiversity: Through their Research, ANTHRA discovered that many important species of plants which are important as medicinal plants or as fodder and many local breeds of livestock are rapidly disappearing. Small pilot projects have been initiated at the local level to serve as demonstration plots. Important species of medicinal and fodder plants are being grown in these plots and propagation methods are being documented. The Animal Health Workers have been simultaneously taught medicine-making and are encouraged to use cultivated plants rather than plants which are found in the wild and are in danger of extinction. Growing awareness of the importance of different plants has created the platform for communities to come together and work towards efforts to save many of these diverse species such as seed banks. These nurseries are becoming mini laboratories where interesting field experiments are being conducted

(contd.)

(Box - 11 contd.)

as well with the help of the village communities. School Children visit these plots to learn more about their immediate environment and the need to preserve it.

Reorienting curricula

To nurture and support research which is people-oriented, the curriculum in Universities has to change so that students are taught more about the history of livestock rearing in the country and the history of veterinary medicine in the country. Curricula should focus on the different systems of livestock rearing, their strengths and drawbacks. Universities in the different agro-ecological zones should specialize in the specific characteristics, breeds, species and systems of their region in relation to the agro-ecological practices within.

The role of different communities as well as the role of women, of children and of traditional healers within different systems have to be introduced to students of veterinary and animal science so that they are better equipped to deal with the problems they are likely to face when they enter the field and are more responsive to the needs of their clients who are different categories of livestock owners.

Caring for the Future

So instead of regretting the fact that there are not enough good dairy breeds like Europe it is more important to examine the good resources we have, where they are located and carefully plan for the future. There are excellent meat breeds of goat, table breeds of poultry, dual purpose breeds of cattle, healthy strains of pigs in remote parts of the country. In fact, the list can be quite extensive if one cares to look.

Thus, if we have to live up to the traditions of our ancient land, the land that preaches *ahimsa,* tolerance, forbearance and love to all without distinction; that which preaches an unswerving kindness towards the members of the animal kingdom, we have to care enough to secure the future for the life forms that live within. We need to care enough to preserve our different livestock resources, the different breeds and the different species. We need to preserve the different food and fodder crops, the grasslands and the forests. We need to ensure that communities who care for these resources, the pastoralists, the people who live in forests are there to help recognize and manage these resources. We need to develop models based on mutual cooperation rather than conflict between different resource users. There is no single model, no one correct formula, which will solve all the problems. The issues are too deep and too complex for that. But as the Buddha has said, 'we will have to work out our own salvation with diligence'.

7

Framework for an Alternate Policy

Framing a policy, even an alternate policy is no easy matter. Ideally, it requires the consensus and participation of a large number of stakeholders. Further, for a livestock policy to be effective, it is equally important that policies governing water, forests, agriculture, land use, energy and environment are effective and in place. Contradictory policies will only defeat each other and negate the effects.

Features of an Alternative Policy

- Planning exercises need to be decentralized so that regional features get consideration. Panchayat raj institutions, Village committees, Forest Management committees, Watershed committees etc, are to be involved in the local level planning exercises.
- Representation of marginalised communities and women in planning exercises.
- Database Management Systems of natural resource to be used for local level planning and decision-making.
- Demand and supply studies need to be initiated at the local level to witness flow of resources within the region. This will further help planning and decision-making.

Capacity Building at Local Levels

For effective decision-making at the village level, the capacities of people here will have to be strengthened.

Local Indigenous Knowledge (I.K) systems to be validated in order that they may be used for capacity building. I.K systems in the fields of animal health, nutrition, management, breeding, housing, forest and grassland management need to be carefully documented, validated and used for training and enhancing the capacities of the local people.

Marginalised communities and women need to be actively included in capacity building and training programmes so that they can participate effectively in village level decision-making.

New Credit Mechanisms at the Village Level

A considerable number of our villagers are still crippled by local money lenders who lend money at extortionate rates. More work will have to done on devising credit mechanisms through which livestock owners can get the necessary credit to purchase and manage the animal of their choice.

Strengthening Local Markets

Many livestock products are easily perishable, for example, milk, eggs, etc. Strong local markets with good transport links ensure that these products reach the market early and in peak condition. This would get better prices for the farmers. Existing local markets, cattle fairs and bazaars can be further strengthened so that they can become focal points for fair exchange of livestock and livestock produce. These bazaars and fairs can also become centres for sharing of information related to livestock care.

Livestock Breeds

Different breeds reflect the economic, cultural and social heritage of different eco-regions and different communities. Genetic diversity also provides the capacity to maintain and enhance animal production, as well as provides the ability to respond to changing conditions.

Studies on the relative merits of indigenous breeds need to be carried out and results shared widely with research institutions, NGOs and civil society groups.

A sound breeding policy has to be developed at the national level in tandem with state level policies.

Careful rules have to be framed to govern all breeding technologies and methods such as castration, introduction of new breeds, artificial insemination, embryo transfer, cloning etc. The uncontrolled or indiscriminate use of any of these techniques can have far-reaching undesirable consequences. Organisations and institutions employing these tools should be governed by these rules.

Strengthening of Local Breeds, Breeders and Breeding Systems

Livestock Census reports should have details of indigenous breed populations. Livestock reports and census to be made available as early as possible to make decision-making effective.

Developing Fodder Resources

Productive livestock is dependant on good fodder. The different fodder resources, agricultural land, grasslands and grazing lands, wastelands, watershed areas and forests will have to further developed. These initiatives will have to be built up from the village level after appropriate estimates are made at the local level of demand and supply of fodder. Presently, while supply of fodder is a state responsibility, under the Animal husbandry Department, grasslands are the responsibility of village panchayats. This leads to a conflict of interests and needs to be changed.

Village grasslands remain the property of a village community and there must be well designed legislations which govern its disposal. This would ensure that these lands are not indiscriminately sold, appropriated by private users or destroyed. Legislation is needed which prevent efforts to sell, appropriate for other purposes or destroy village grazing lands. Local biodiversity related to fodder species need to be carefully studied. Indigenous Knowledge related to fodder species, management of fodder resources, the preservation of fodder should be taken into consideration while planning for fodder development especially in watershed areas and joint forest management areas.

Grazing policies need to be established for common grazing areas; these maybe water shed areas, forest areas, and grazing lands in consultation with all stakeholders. Traditional grazing rules like those for the Orans in Rajasthan ought to be explored.

Planning for Disasters

Appropriate planning exercises will have to be undertaken to plan for disasters such as droughts, cyclones, earthquakes etc. Local communities need to be involved in the planning exercises so that when the disaster does strike they are equipped to deal with the immediate crisis.

Animal Health and Management

Animal disease leads to poor production and poor quality of produce. This may be countered through the following aspects:

- Thrust to be laid on preventive health.
- Improved management practices prevent many diseases. Alternative systems which integrate the best of modern and indigenous knowledge suitable for different eco-regions developed and shared with livestock owners.
- Preventive calendar of management practices to be developed at different levels and steps taken to ensure availability of services at critical periods.
- Essential vaccines for all categories of livestock to be made available by the state in adequate quantities.
- Vaccines for backyard poultry, goats, pigs and other livestock kept by marginalised communities to be made available.
- Infrastructure for distribution and delivery of vaccines to livestock owners at the village level ought to be developed.
- Opportunities for manufacture of thermo stable vaccines to be explored.
- Training and capacity building of para-professionals to deliver vaccines and other preventives at the village level.
- Prevention and eradication of certain fatal, communicable and zoonotic diseases to be undertaken on a war footing.
- Alternative systems of medicine such as indigenous systems, folk medicine, ayurveda and homeopathy to be explored and integrated suitably into the existing system, as this will help reduce the cost of expensive medicines and chemo-therapeutic agents.
- Training of para-professionals to deliver primary veterinary health-care at the village level.

Research

The following areas of research needs to be explored:

- Research to be small farmer, livelihood and environment-oriented and not exclusively market driven.
- Research on indigenous systems of breeding, feeding, management, grazing, health, fodder, feed additives, medicine to be undertaken to identify valuable traits, production parameters and suitability for specific agro-eco regions so that they can be successfully integrated into different programmes as well as used for capacity building.
- Research on integrating different systems of medicine.
- Studies on livelihood patterns and methods of strengthening them especially with reference to pastoral groups and tribes.
- Studies on the effects of agro-chemicals and pesticides and the effects when they are consumed on animal and human health as well. Potential markets especially markets dealing with organic food are also lost as a result of the indiscriminate use of pesticides.
- Research to find suitable alternatives to chemical pesticides and chemical feed additives.
- Research on emerging concerns such as livestock environment interactions.
- Research studies and information ought to be widely circulated so that farmers can make an informed choice.

Curriculum in Veterinary Colleges

- Client-oriented and small farmer-oriented.
- Livelihood and environment concerns of livestock rearing should become a part of the undergraduate curriculum.
- Different colleges in different agro-ecological regions should specialize in livelihood systems of their region.
- Course curriculum should include reference to local knowledge systems, their benefits and the successful integration of modern and traditional systems.

References

ACCM International Private Ltd, Alpha Agritech Consultants, Bangalore and JPS Associates India. (1994). *Livestock Sector Review Executive summary. Prepared for World bank and Government of India.*

Agarwal, A., Narain, S. (Eds.). (1985). *The State of India's Environment 1984-85, the Second Citizen's Report*. Delhi: CSE.

Alvares, C. (Ed.). (1999). *The Organic Farming Source Book*. Goa : The Other India Press.

Anjaria, J. (1984). *Traditional (Indigenous) Systems of Veterinary medicine for Small Farmers in India*. Bangkok: FAO.

ANTHRA. (1997). *Community-based research on local knowledge systems: The ANTHRA project on ethnoveterinary research in Ethnoveterinary Medicine Alternatives for livestock development*: Proceedings of an International Conference, Pune.

ANTHRA. (1998). *Internal report Summary report of IKAH Project, July 1998 – June 1999*. Secunderabad and Pune: ANTHRA.

ANTHRA and Girijan Deepika. (2000). *An Integrated approach to Reducing Production losses in Backyard Poultry Production Systems in Tribal Areas of East Godavari district Andhra Pradesh*. Secunderabad: ANTHRA.

ANTHRA. (2000). *Report of training programme held in Village Ginabahar, district Sarguja, M.P submitted to CRS Bombay.*

Bisio, A., Boots, S. (Eds.).(1995). *Energy Technology and the Environment* (Vol I). Canada: John Wiley & Sons.

Braudel, F. (1979). The Structure of Everyday Life. In *Civilisation and Capitalism 15th –18th Century* (Vol I). London: Collins.

CED Research Team. (1982). *Operation Flood, Development or Dependence*. Bombay.

Cincotta, R., Pangare, G. (Eds.). (1993). *Pastoralism and Pastoral Migration in Gujarat*. Anand: Institute of Rural Management.

Delgado, C., et al. (1999). Livestock to 2020: The Next Food Revolution: Food, Agriculture and the Environment. Discussion paper No 28. Washington DC: International Food Policy Research Institute, Food and Agriculture Organization of the UN, International Livestock Research Institute, Kenya.

The Department of Science and Technology, TIFAC. (1996). *Agro Food processing: Technology Vision 2020*. New Delhi: Government of India.

Dharampal. (1983). *Indian Science and Technology in the Eighteenth Century: Some Contemporary European Accounts*. Hyderabad: Academy of Gandhian Studies.

Girijan Deepika, ANTHRA and Yakshi. (2001). Community efforts of in situ conservation of the Aseel Poultry Breed in the Adivasi areas of East Godavari district. Proceedings of an international Conference on Conservation of Indigenous Livestock Breeds, LPPS, Sadri, Rajasthan, Nov 2001.

Government of Andhra Pradesh. (2000a). *Swarna Andhra Pradesh Vision 2020*. Working paper on Livestock development Policy. Department of Animal Husbandry: Government of Andhra Pradesh.

Government of Andhra Pradesh. (2000b). *Swarna Andhra Pradesh Vision 2020*. Working paper on Agriculture Development Policy. Andhra Pradesh: Department of Agriculture.

Government of India. (1972). *Report of the Irrigation Commission* (Vol. 2). New Delhi: Ministry of Irrigation and Power.

Groombridge, B.(Ed.). (1992). *Global Biodiversity: Status of the Earth's Living resources: Compiled by World Conservation Monitoring Centre*. London: Chapman and Hall.

Gupta, K.A. (1992). "Integration of livestock with agro-climatic zone–based land use planning". *Livestock Research for rural development*, 4 (1).

Hiremath, S.R., Kanwalli, S., Kulkarni, S.R., (Eds.). (1995). *All About Draft Forest Bill and Forest Lands*. (3rd ed.). Pune, Dharwad: Samaj Parivartan Samudaya and Center for Tribal Conscientisation.

Horowitz, M., Jowkar, F., (1992). *Pastoral Women and Change in Africa, The Middle East and Central Asia*. Institute for Development Anthropology. Working paper No 91. Binghamton, New York: IDA.

Howard, A. (1956). *An Agricultural Testament*. Mapusa: Other India Press.

Indo Swiss Goat Project, Ajmer. (1993). *Manual for Improved Goat production*.(Parts 1-4). Berne: Intercooperation.

Jodha N.S. (1995). *Sustainable Development in Fragile Environments: An operational Framework for Arid, Semi arid and Mountain Areas*. Ahmedabad: Centre for Environment Education.

Kalam, A.P.J., Rajan, Y. S. (1998). India 2020: A Vision for the New Millennium. New Delhi: Viking (Penguin).

Kautilya. *The Arthashastra*. (L.N. Rangarajan, Trans,). New Delhi: Penguin Classics.

Kavoori, P . S. (1999). *Pastoralism in Expansion: The Transhuming Sheep herders of West Rajasthan.* New Delhi : Oxford University Press.

Khilnani, S. (1998). *The Idea of India*. London: Penguin.

Korten, D. C. (1998). *When Corporations rule the World*. Mapusa: Other India Press.

Kothari, A., Singh, N., Suri, S.(Eds.). (1996). *People And Protected Areas: Towards Participatory Conservation In India*. New Delhi: Sage.

National institute of Science Communication. (1998). *The Wealth of India: A Dictionary of Indian Raw Materials and Industrial Products* (Vol. 6). New Delhi: CSIR.

Kurup, MPG. (1999). *National Livestock Policy in India*. In V. Ahuja (ed.), Workshop on Commercialization of Livestock Health and Breeding Services: Papers and proceedings of Indian Institute of Management Ahmedabad, the World bank and the Swiss Agency for Development and Cooperation, Berne.

Moorhouse, G. (1985). Paladin: India Britannica.

Nell, A. J. (1998). *Livestock and the Environment*. Proceedings of the International Conference on Livestock and the Environment, 16 –20 June 1997, International Agricultural Centre, Wageningen, Netherlands.

Mc Neill, J. R. (2001). *Something New Under the Sun: An Environmental History of The Twentieth Century.* New Delhi: Penguin.

Pagot, J. (1992). *Animal Production in the Tropics and Sub tropics*. Wageningen, London: Technical center for Agricultural and Rural Cooperation, and Macmillan Education Limited.

Pereira, W. (1993). *Tending the Earth: Traditional Sustainable Agriculture in India*. Bombay: Earth Care Books.

Pillai, G.M. (Ed.). 1999. *Challenges of Agriculture in the 21st Century*. Pune: MCAER.

Qazim, S.Z. (1996). *National Livestock Policy perspective*. New Delhi: Planning Commission, Government of India.

Ramdas, S., Ghotge, N. (1995). *An Analytical Study of a Community Based Approach to Sheep , Development Implemented by India Development Service, In Partnership with The Semi Migratory Shepherds Of Dharwad District, Karnataka*. Unpublished, ANTHRA, Secunderabad.

Ramdas, S., Ghotge, N. (1996). *Ethno veterinary research in India: An Annotated Bibliography*. Unpublished, ANTHRA, Secunderabad.

Ramdas, S., Ghotge, N. (2001). Vision 2020 Livestock Development Policy of Andhra Pradesh India: A skewed vision of food security for Indian farmers. In Gabrielle Stoll(ed), *Drawing on farmers experiences in food security: Local successes global failures.* Bonn: German NGO Forum on Environment and Development.

Randhawa, M.S. (1959). *Farmers of India* (Vol 1 and 2). New Delhi: ICAR.

Rathore, H. S., Rathore, S. S., Rollefsson, I. K. (1997). *Traditional animal health services: a case study from Godwar area of Rajasthan in Ethnoveterinary medicine Alternatives for livestock development*. Proceedings of an International Conference on Ethnoveterinary medicine, BAIF Pune, 1997.

Rifkin, J. (1995). *The End of Work: The Decline of the Labour force and the Dawn of the Post market era.* New York: G.P Putnam & Sons.

Rollefson, I. K., Rathore, H. S. (1999). *Camel Breeding in India: A Status Report with Policy Recommendations*. Sadri: Lokhit Pashu Palak Sansthan.

Sahai, R. and Vijh, R.K. (Eds.). 2000. *Domestic Animal Diversity Conservation and Sustainable Development*. Karnal: SI Publications.

Shirlaw, L. (1940). A short history of Ayurvedic veterinary literature. *Journal of Indian Veterinary Science and Animal husbandry* Vol10(1): Page 1-39.

Singh, H. (1994). *Domestic Animals*. New Delhi: National Book Trust.

Steinfeld, H., Haan, C., and Blackburn, H. Livestock – Environment Interactions, Issues and Options: Report of a study conducted for the European Commission Directorate, FAO, World Bank.

Steinfeld, H., Haan C., and Blackburn, H. Livestock – Environment Interactions, Meeting the Challenge, European Commission Directorate FAO World Bank.

Tata, S.N., et al. (1990). *Handbook of Animal Husbandry*. Delhi: ICAR.

Vijaylakshmi, K., Shyam Sundar, K.M. (1994). *Mrgayurveda*. Indian Animal Science LSPSS Monograph No 13 and 14. Madras: Lok Swasthya Parampara Samvardhan Samithi.

Watt, G. (1972). *A Dictionary Of The Economic Products Of India* (Vol. 3). Dehradun : Bishen Singh, Mahendra Pal Singh.

Index

D

E

F

G

H

I